Author, 1935

The author being fired up in Heinkel: Ted Gibson, Author's wife, Arthur and Geordie.

Stainforth and the S6B in 1931

Frierson, Green, Birch and Shepherd

Macarka

207 Squadron, 1937

ROLLS-ROYCE
FROM THE WINGS
Military Aviation 1925–71

RONALD W. HARKER

ACKNOWLEDGEMENTS

I should like to thank my family, Charles Cain, David Birch, David Dorrell, Ken Smith and Dennis Miller-Williams for their many helpful suggestions and friendly criticisms and also Sir Christopher Foxley-Norris for agreeing to write the Foreword. I should also like to thank Jane Marshall the editor, who was so helpful in providing the continuity of my book,

PHOTOGRAPHS

I should like to thank Rolls-Royce for their many photos; *Air Pictorial* for the photos which appear on pp. 103, 104, 105, 106, 107, 108; *Flight*: p. 110; Ken Smith: p. 113; Stuart Howe: p. 49, and top left picture on the cover.

Oxford Illustrated Press, Shelley Close, Risinghurst, Oxford
Made in Great Britain by Morrison & Gibb Ltd, Edinburgh, Britain
ISBN 0 902280 38 4

Contents

Foreword

by Air Chief Marshal Sir Christopher Foxley-Norris,
GCB, DSO, OBE, MA.

This is essentially a personal story but nevertheless it recounts matters of great national and international interest. It tells from a uniquely informed position the history of forty years and more of the great firm, indeed industry, of Rolls-Royce during a time of unparalleled advance in technology and performance; and shows the emergence of aviation, both military and civil, as we know it today.

The name of Ronnie Harker is familiar to two full generations of those in that industry but perhaps even more notably to those in the Royal Air Force and the other airforces with which Rolls-Royce and the RAF have been intimately associated. He rose from a humble (reasonably so) apprentice, straight from public school to the critical appointment of Military Aviation Adviser to the firm; and his career spanned the vital periods of pre-war re-armament, war-time development, and post-war advance into the supersonic era. This was a time when aircraft, their engines and their equipment advanced far more remarkably and rapidly than in their comparatively inert first thirty years. Ronnie Harker as pilot, engineer, negotiator and promoter of projects played a vital part in ensuring that both his firm and his country kept pace with not only our wartime enemies but, perhaps just as decisively, our peace-time friends.

But it is not only for his professional performance that Ronnie will be remembered. The link between industry and armed forces is fundamental to the interests of both; in the years here described it was a firm and close link, whose importance was mutually appreciated and which depended largely on personalities. Ronnie was known, respected, trusted and liked throughout the Royal Air Force at all levels from the flight-line to the Air Council room. When, as regrettably now seems to be the case, such links are allowed to become weaker and looser, both parties suffer and the security of our country with them. It is good to know he continues to be active as an aviation consultant, thus enabling him to maintain close contact with his former colleagues and friends.

Told in a light-hearted, self-effacing style wholly typical of the man himself, this story nevertheless describes matters of historical interest to us all and at the same time of considerable bearing on our present position and future prospects. I have found great pleasure in reading it and I am quite sure that many others will do the same.

Introduction

Rolls-Royce. What a glorious name! How it rolls off the tongue—especially as Mrs Holdsworth used to say it when she was on the telephone exchange at Conduit Street. It epitomises the spirit of excellence which has been built up since Royce met Rolls and the pattern was set.

In achieving this unique reputation, many admirable qualities came into being: integrity, consideration for the customer, close relationship with the men on the shop floor, a team spirit combined with the satisfaction of doing a job well and a sense of humour. These coupled with hard work were the ingredients which made up a recipe for success. A tradition of loyalty was built up amongst all levels of staff and work's personnel which was unique in industry. Labour relations were very good; there were families who had three generations serving at the same time. Promotion was on merit and there were opportunities for all. There was a recent example of this special loyalty in 1971, when over a million workers came out on strike against the Industrial Relations Bill, but 98 per cent of the Derby workers went to work as usual, thus defying the unions.

There was a family influence within the company which stemmed from the Hives family. They gave Christmas parties and garden parties in the summer at Duffield Bank Guest House. To be invited to these was considered a great privilege; they were great and enjoyable occasions. There was a certain feudal air about them; a modern way of thanking the tenants for their labours. The main thing however was that they furthered and strengthened the team spirit which was so important in Rolls-Royce. Only very rarely was any individual singled out for praise; it was the result that mattered and the collective effort which achieved it.

Ernest Hives became prominent in the company during the First World War when Rolls-Royce was just beginning to gain its great reputation in aviation. It was he who raised the company to its peak by his great leadership and personality. He brought together the team which produced the R engine and the Merlin engine, and he also encouraged the close association which formed in the Second World War between the company, the Royal Air Force, the Royal Navy and the Government. The greatness of the company was really founded on these cornerstones together with the principles of engineering excellence which had already been laid down by Sir Henry Royce earlier. His motto was 'whatever is rightly done, no matter how humble is noble' and this was always the guiding principle. After the war however the jet age arrived and this marked changes much wider than those just in engine development.

The first most noticeable one was the rapid development of civil aviation, the engines for which at first rode on the back of the military but were soon competing with us on the military side for development time and funds. It was also discernible about this time that a profit consciousness crept into the company and a more commercial attitude was developed. At the same time it was benefitting from a string of successes and with a new generation of young brilliant men coming in, an attitude of arrogance was forming which was noticeable in our dealings with the Ministries and the aircraft constructors. This was soon nipped in the bud by the senior staff who decreed that a more humble approach should be taken. Later, when crises arose and services became acute, this humble approach plus the ability of the company to stop trouble and rectify faults quickly, paid off handsomely.

All the while, prosperity increased and the firm grew in size; there were 4,000 people when I joined in 1925 and there were 80,000 when I left in 1971. There is no doubt that as the size of the company increased, the management and the worker relationship became less personal. There were many more meetings with many more people involved and so it became much more difficult to just have a friendly chat with someone during which one could convey any customer's criticisms and where things maybe going wrong. A time lag began to creep in between receipt of a complaint and action to effect a cure and a slowing down process was taking place at a time when time itself was at a premium. This was a problem of management.

One of the most important contributory factors in the company's downfall was the ambition of Rolls-Royce to challenge the USA engine builders on their own ground by competing for the contract for the Lockheed Tristar with the RB–211 engine. This was to be proof that Rolls-Royce was ahead of her rivals on technology. Alas, the financial arrangements could not be met—inflation and a world recession in civil aviation saw to that—and so the Collosus fell.

The great reputation of Rolls-Royce and the fact that it is now synonymous with integrity, engineering excellence, technical achievement and respect for the customer is something that is almost taken for granted today and it is with a feeling of great pleasure and pride that I realise that I and my many friends and colleagues were there in her formative years. It was always a matter of great pride to know one was representing Rolls-Royce; to answer the phone and be able to say 'This is Harker, Rolls-Royce speaking' was a thrill even to the end.

I was prompted to write this book mainly because I felt that I owed a debt of gratitude to the company for all the enjoyable and interesting experiences I had while I was with them. Secondly, some of the controversial projects with which one was involved I believe should be recorded in perhaps a more revealing light than the official histories have been able to do. Working for the company was really a full-time job; we were dedicated and the values for which the company stood certainly permeated into ones own life. I met many colourful people not only within the company but in the Services, Industry and in many parts of the world, and as many of these people became close personal friends, so the gap between one's life at work and at home narrowed even more. Somebody once said 'You get Rolls-Royce stamped on your backside'; for me it was more than that, it became a way of life.

The early days of the company, and in particular the development of the cars, have been well recorded. It is the hope and intention of this book, which covers the years between 1925 and 1971 that it will help redress the balance by showing the development of aviation, and military aviation in particular in this time.

DEDICATED TO

Those special people who have by their
inspiration and example been an influence
and encouragement at all times.

Early Days 1

I joined Rolls-Royce as one of the annual intake of twelve premium apprentices in 1925. To begin as a premium was to start from a privileged position: one's father paid the firm £400 and in return for this we were apprenticed to the firm for four years during which time we were to work in all the departments. At the end of this period we were expected to understand the whole process of the manufacture of cars and aero engines.

In joining Rolls-Royce, I was following my brother who was nearly three years older than me. When he had told our father that he wanted to go into motor engineering and join Sunbeams (who had just won the Grand Prix), he had said 'If you must go into motor engineering, there is only one firm, the best, and that is Rolls-Royce'. Little did we know how right he was, and I am forever grateful for his wisdom. He was terribly proud of our joining the firm and a few years later, when introducing us to some of his doctor friends he said 'the boys are working with Rolls-Royce you know; Rolls-Royce make the engines for the RAF, and Hawkers and people make the wings and things'. This story was rarely appreciated when told in later years at the tables of the various plane-makers!

It was a big change in life to go from my public school in Shrewsbury to living in digs in Derby. The early days of going to work are very clear in my mind. We used to have to be at work by five minutes to eight and I used to rush off to the works on my bicycle after a hurried breakfast in order to cover the two miles in time. There were two hooters known as 'bulls'; one blew at ten minutes to eight, the second at five to eight, then the gates closed and if one didn't get there in time one was locked out until after lunch. Most of us got caught at one time or another! On arrival we had to clock in, change into our overalls and start work. A report on attendance was submitted to the Works Manager, Mr Wormald and this was carefully copied and sent to one's father; just like school.

As premiums we changed from one department to another each month, beginning in the foundries and gradually working our way through the fitting and machine shops until we reached the test department. This was the most interesting as there was a great variety of processes to be studied. The company made everything on the chassis except batteries, sparking plugs and magnetos. All other electrical equipment, including bakelite mouldings, switches, starters

The Rolls-Royce Works at Derby in 1925

and dynamos were made at Derby and were of a higher quality than contemporary equipment on the market. We learnt the meaning of quality as we went through the shops and how parts were scrapped if they were not within the close limits demanded.

Getting to know the workers was also a valuable experience, and we soon came to understand the importance of the management-worker relationship. They varied a lot in their reaction to us: the majority were easy to get on with and very willing to pass on their knowledge and practical skills, others resented the privileges of the premiums and so were less forthcoming.

One of the privileges the premiums enjoyed was the use of a loo which had a door that locked. We used to collect the key from Mr Asbury who sat at the seat of custom and kept a check on the time taken by each of us. On the inside of our special loo, written on the wall was the never to be forgotten advice: 'Its no use standing on the seat, Rolls-Royce crabs can jump six feet!' One was subject to many other examples of rough workshop wit, all part of the process of growing up.

It is worthwhile here describing the routine testing that each chassis went through. First of all, after initial assembly, the engine was run in on coal gas for six hours; components such as dynamos and starters had previously been run in and tested for performance on rigs. The next test was a series of power curves on the brake and a preliminary check on the engine and gearbox for noise; these tests took about six hours. The engine was then installed on a chassis which had already been allotted to a customer. The chassis then went through a dynamometer test where the power was recorded through the rear wheels, thus testing the gearbox and back axle. Following this, the chassis was kitted up for the road and was taken on a fifty mile road test known as 'stage five'. This test would show up a noisy back axle or gearbox and if one was discovered then the tester had to persuade the foreman of the department concerned to re-bed the gears. This was usually quite a difficult job as it would affect the department's weekly output and hence the bonus!

After stage five test, the engine was decarbonised, tuned and got ready for the first 'final'. This consisted of a twenty-five mile run, including timed climbs on local test hills. If this test was completed satisfactorily the engine was finally checked and offered to the chief tester for its final. If the chassis was passed by him it went to 'quality' where the quality tester tried it; he represented the inspection department. The chassis would then be despatched, if passed as up to standard, by rail to London to one of the coach-builders to have the body fitted. The complete car was then finally tested before being handed over to the customer. The cars were guaranteed for three years, they cost around £3,000 and the output varied between twenty-five and thirty cars per week.

There was little real activity on the aero side between 1925–9; the company was still making the World War 1 engines such as the Eagle, Falcon and the Condor. We worked on these both in the fitting shop and on the test

Rolls-Royce premium apprentices in 1927

beds. During 1927 however we did notice that there was a strange new noise coming from the test beds. It was a sharp crackle rather than the muted boom of the Eagles, etc. which we were used to. One of the test beds had been taken over by the experimental department and it was shrouded in secrecy. By snooping around, however, we did see that it was quite a small engine and that it was all aluminium—quite different to the older types of engine we were used to seeing and working with. We later found out that this was the new F-10 engine.

Our apprenticeship at Rolls-Royce terminated with a month in the car test department and this was most enjoyable. The testers were very decent fellows, and I got to know Jack Pepper, Doug Fox and Tommy Bryan particularly well. We used to drive all over Derbyshire in all weathers and they were kind enough to invite me into their homes on occasions. The cars I tested included the Silver Ghost, the Phantom 1 and the Phantom 2.

The author taking a Phantom I out on test.

On the last day of my apprenticeship I was summoned to the presence of Mr Wormald. He told me that I had had good reports from all departments but that because of the lack of opportunity due to the general slump, he thought I should leave and seek a job elsewhere in the motor industry. I was neither pleased nor flattered and I left his office and went straight along to see Mr Maddox, the manager of the car test department. He kindly agreed to keep me on for awhile and so I stayed on as an assistant tester earning the princely sum of £3 18s. a week. I was pleased to be earning my own living for the first time.

After my first four years at Rolls-Royce, I had only really learned to be a practical engineer, although by attending the Derby Technical College and passing the Institute of Automobile Engineers exam, I had learned a certain amount of theory and design. I was very much aware that I still had no inkling of the sources of energy which made Rolls-Royce tick. This gradually came later as one was given more responsibility and the opportunity of associating with more senior people. To grow up in Rolls-Royce was a privilege which one did not fully appreciate at the time, although there was always the feeling of confidence that one was working for the best and most famous engineering firm in the world.

My job as assistant tester was most interesting and pleasant but in 1930 the economic recession happened and very quickly the works went on short time, and the orders dried up. The firm had always made the cars to customers orders—reading the job cards was like reading a page in *Debret*—and now for the first time, a stock of new cars piled up. A number of we youngsters became redundant and were laid off and so I had no choice but to pack up and go home to Newcastle.

I think my desire to fly probably originated when I went with my father to the Hendon Air Pageant in 1927 at the age of eighteen. The new Rolls-Royce 'F' engine was on public show for the first time flying in the Westland Wizard and the Fairey Fox. The performance of these machines was vastly superior to the standard RAF types, the Bulldogs, Siskins and Gamecocks and the crackle of the twelve-cylinder stub exhausts and the smell of the Castrol 'R' of these new engines was thrilling. I was greatly impressed and so was born my desire to become a pilot.

Of course at this time it was a very passive desire; it was beyond my wildest dreams to think that I would ever fly a Fury, Hart or Fox—my imagination only stretched to Moths or Avians! But it shows how the apparently impossible can come to pass and even be surpassed, for exactly ten years later to the day, I was not only flying Furies and Harts and planes with even higher-powered experimental engines, but I was also taking part in the 1937 Hendon Air Pageant as a member of 504 County of Nottingham Auxiliary Squadron!

But it was now only 1930 and I was back at home and out of work. I was finding it very difficult to find a job. I had written off to many firms and individuals but had had no luck. Two of the people whom I contacted were W. O. Bentley and Captain George Eyston. Bentleys were no better placed than Rolls-Royce to take on staff and Captain George Eyston, who was at that time one of the leading racing drivers breaking speed records all over the world, was very sympathetic and understanding but again could not help. It was then that I began to think about learning to fly. I thought about joining the RAF on a short service commission but was dissuaded from applying by my father who wanted me to learn flying locally in the hope of getting back to Rolls-Royce when times were better. I therefore joined the Newcastle Aero Club at Cramlington.

Learning to fly there was most enjoyable. My instructor Mr McGevor was most patient and helpful and I soon progressed under his guidance—perhaps too quickly though—one day I flew over Gosforth Park Racecourse and got back to find that some complaints had been made about my flying too low, although I had not flown below 2,000 feet, and I was suspended for two weeks!

It was necessary to join another club in order to complete my full flying course and so I went to Lympne and joined the Cinque Ports Flying Club. It was while I was here that a letter arrived from Cyril Lovesey, who was in charge of the aero engine side of the experimental department at Rolls-Royce, asking me to return to the company to take part in the development of the 'R' engine for the Schneider Trophy. I was absolutely delighted and immediately went to Derby to see both Cyril Lovesey and Hives; they told me to report for work in a week's time. What a great relief this was! The problem of trying to find a job was over and I could once more see a career unfolding before me.

The history of the 'R' engine is significant because it was developed from the 'F' engine upon which the whole future success of the Rolls-Royce aero engine division was built. Air Marshal Sir Geoffrey Salmond who at the time was Air Member for Supply at the Air Ministry, told me the story of its development.

It began with Sir Richard Fairey, head of Fairey Aviation Company, who went to the States to see the Curtiss D-12 engine. This engine was of advanced design embodying wet-cylinder liners in an aluminium block instead of separate steel cylinders with welded water jackets as used in the old war-time Rolls-Royce engines. The Curtiss engine was a twelve-cylinder vee, very light and with a small frontal area.

Sir Richard saw that if he built this engine and installed it in his own aircraft instead of using Rolls-Royce engines or the Napier Lion, he would be in advance of his British rivals. He successfully negotiated the licence and began building the engine in Britain; it was called the Fairey Felix, and the first plane to have one installed was a light bomber called the Fairey Fox. This went into service with the RAF and was first introduced into 12 Squadron based at Andover. These were the aircraft which had so impressed me at Hendon with their superior performance and thrilling exhaust noise.

The Air Minister saw the possibilities of this modern engine installation and as there were doubts over the ability of an aircraft company to produce engines. It was decided that Rolls-Royce should take over the engine for experimental tests and evaluation. An engine was therefore sent to Derby for examination. This was an unusual move; it was normal practice in the motor car division where there were always a number of foreign cars under test in the experimental department (such as a Lancia, Lambda, Graham Page, Packard and a Hispano Suiza, data which had enabled Rolls-Royce to develop front wheel brakes), but it had not been done with aero engines before.

The Curtiss D-12 was examined and tested and it was decided to design a completely new engine, based on the D-12 but incorporating well-proven Rolls-Royce features. The design team was led by A. J. Rowledge and Elliot under the aegis of Sir Henry Royce. Rowledge had moved from Napiers where he had been in charge of the famous Napier Lion. The result of their work was the 'F' engine, which was later called the Kestrel. Rather sadly, however, the RAF finally awarded the contract for a light bomber using the Kestrel to Hawkers; this was a big disappointment for Faireys who had done the pioneering work on these new sharp-nosed aircraft. However Faireys did land a large order in Belgium both for the Fox and the Firefly with the Kestrel engine. The new Hawker light bomber was called the Hart and was the first of a famous family of aircraft of which many variants were made. After these installations, a centrifugal supercharger was fitted, and the new improved engine was installed in the Hawker Fury, the Fairey Firefly, the Westland Wizard, and the Avro Antelope. The supercharger work was done by Jimmy Ellor, who had recently joined the company from the Royal Aircraft Establishment (RAE), and it was his development work which was to enable the engine to always keep ahead of its competitors.

Rolls-Royce enjoyed tremendous success with the Kestrel engine: it was soon installed in the Audax for Army co-operation, the Hind as a day bomber, the Osprey as a carrier-borne bomber, the Demon as a two-seater fighter, the Nimrod as a fleet fighter and the Hawker Fury, the finest of them all, as an interceptor. I flew many hours in the High Speed Fury which was used for installation development at Hucknall, and it was my favourite aeroplane. These

The High Speed Fury

aircraft were subcontracted to Vickers, Armstrong and Whitworth and Glosters, many being built and exported to foreign countries. It was specified also for the new bombers under development, such as the Handley Page Heyford, the Fairey monoplane bomber, then the Vickers twin- and four-engined bombers, the Short flying boats and a number of foreign aircraft too, e.g. the Fokker VIII, a Dornier bomber and a number of others. Its rivals were the Bristol air-cooled radial engines, the Pegasus and Mercury, and good though these were, the bulk of the orders went to Rolls-Royce. However, competition was good and it helped to improve the breed. The Kestrel must go down in history as the most successful aero engine between the wars and the forerunner of the

Merlin which followed and enabled the Allies to dominate the air war in all the various theatres. Thus the decision of Royce, Elliott, Rowledge and Hives to abandon the old design and embrace the new was so amply vindicated.

Fairey still had the ambition to break into the aero-engine business and he obtained the services of Major Forsyth from the Air Ministry, where he had been in charge of engine development. He designed an ambitious range of engines, a twelve-cylinder vee engine known as the P-12 Prince, was followed by a sixteen-cylinder engine and later by a P-24 Prince of twenty-four cylinders. These were intended to compete with Rolls-Royce and were flown in the Fairey Battle. Although showing considerable promise they were not supported by the Air Ministry and eventually they were discontinued. This was unfortunate in some ways as they were advanced in concept, and in conjunction with the experience of Fairey in building propellers, they might have shown a general improvement in performance when used in conjunction with the new generation of monoplane fighters now on the drawing board. They would also have provided a spur to competition with Rolls on liquid-cooled engines.

At about this time Rolls-Royce had been persuaded to become involved in the airframe business, and a stake in the equity of Phillips & Powis had been purchased. This proved to be an error of judgement as the rest of the aircraft companies objected to it as they felt that favoured treatment would be given to Phillips & Powis and aerodynamic information derived from their own developments might be passed on through the association of Rolls-Royce. They became reticent in passing on performance information. The story goes that one day when Hives was visiting Faireys accompanied by the Managing Director, Sir Arthur Sidgreaves, Sir Richard Fairey remarked, 'We don't like Rolls-Royce getting into the aeroplane-making business', to which Hives replied, 'You need not worry, we treat our partner just as we treat your involvement in aero engines.' 'And how is that?' asked Sir Richard. 'As a ruddy joke!' retorted Hives. There was much competition at this time in the industry and not much money available for development thus generating sensitivity between many of the companies. Rolls-Royce and Hawkers always were closer together than Faireys. Many attempts have been made to combine the manufacture of engines and airframes in one company but never with much success except perhaps in the case of the Bristol company, which has made the famous air-cooled radial engines and also some successful aeroplanes.

The 'R' engine then, was a development of the Kestrel and was a more highly rated and stressed engine designed for a short racing life. This engine powered the Supermarine S.6. which won the 1929 Schneider Trophy. It had delivered 1,800 bhp then and it now had to be further developed if it was to win again in 1931.

Before I begin on my new career with Rolls-Royce and the work on the 1931 Schneider Trophy, one other significant thing happened to me in 1931. When I returned to Derby, one of the first things I did was to join the Leicester Aero Club so that I could continue my flying practice. The lessons were rather expensive: I was paying two pounds an hour out of my weekly pay which then only amounted to four pounds! But I managed to get some practice in at the weekends.

One of my friends at Rolls-Royce at that time was Michael Dawney. He was also keen on flying, and I found that he had joined the Royal Air Force Special Reserve and so not only was he being taught to fly at Hucknall on RAF aircraft, but he was also being paid for it! He took me over to Hucknall where number 504 County of Nottingham Squadron was based and introduced me to the Commanding Officer, Squadron Leader Charles Tor Anderson, and Flight Lieutenant Tubby Dawson. I explained that I would like to join the squadron if there was a vacancy and they were prepared to have me. I was invited to a guest night to meet the other officers. I found the atmosphere friendly and the prospect of joining the RAF on a voluntary basis most exciting. It meant that I would be expected to attend the squadron at weekends for flying tuition and lectures, take part in parades, social activities and the annual fortnight's training camp. When the squadron was fully trained I would be expected to take part in the RAF annual defence exercises.

I was accepted and so it was arranged for me to be flown to Hendon to attend the Central Medical Board in London. It was all very thrilling, particularly when Noel Capper flew me there in one of the squadron's Hawker Horsley Bombers! I quite often see him today at Prestwick where he recently retired from being Chief Test Pilot to Scottish Aviation, a company with which I am associated as a consultant. Having passed the medical examination, I was given a cheque with which to purchase my uniform: it was a proud moment when I first reported to the squadron in the uniform of a pilot officer to begin my flying instruction. Tubby Dawson told me to forget that I already had an 'A' licence and that I would have to start right from the beginning and learn again RAF fashion. He taught me first on an Avro 504-N Lynx and then on a Horsley bomber which had a Rolls Condor engine. I was awarded my wings and became an operational pilot.

The special Reserve Squadrons, of which there were five, numbering from 500 to 504 were based in widely dispersed locations and were intended to supplement the regular Air Force by training volunteers who could take part in peace time exercises and rapidly become proficient in the event of an emergency or a war. The highlight of our year was the annual camp at Hawkinge. At the 1933 camp a rather exciting incident took place which is worth recording.

Towards the end of our stay, we were to be inspected by the Chief of Air Staff, Sir Edward Ellington; the parade over we were to take to the air in squadron formation of nine Horsleys (three flights of three aircraft). It was a hot day with a fickle breeze which kept changing direction. Our display had gone well and the squadron came in to land in separate flights: 'A' flight first followed by 'C' flight. But quite suddenly the wind changed direction 180 degrees and as we taxied in we could see that 'B' flight had not noticed and that they were going to land down wind. Luckily they realised just in time and all three opened up to full throttle to try and go round again. The two wing men were able to remain airborne and to miss the hangars, but Freddie Hartridge hit the

504 Squadron in 1933

roof fair and square and lodged there.

We all got out of our aircraft and rushed to the hangar to try and rescue Freddie and his rear gunner, who were unable to get off the roof. By this time a fire had started and the rear gunner was injured and trapped. The fire tender arrived but only foam bubbled out of the end of the pipe and it was useless. After an agonising few moments the air gunner managed to free himself and fell off the roof escaping with minor injuries. The fire meanwhile had spread to the fuel tanks and Freddie had dropped off the roof and had landed inside the hangar. Being August Bank Holiday, the hangar was locked and the key could not be found—the hangar was now ablaze with Freddie inside! We began to attack the doors and managed to force them open. The squadron doctor, Ronnie Varten, rushed to Freddie's aid and bravely dragged him clear. The hangar was full of old Blackburn Darts and by the time the main doors were opened, the place was an inferno and only one was saved. So ended our inspection by the Chief of Air Staff who fortunately had left Hawkinge just before all this had happened.

This decision to join the Special Reserve Squadron, or the Auxiliary Air Force as it became known in 1936 was to prove to be most important in the shaping of my future career: when I began to specialise in the military aviation, it brought me into close contact with contemporary officers of the regular Air Force where I was accepted not only as a Rolls-Royce representative but also as a brother officer. This was invaluable.

So that 1931 which had started off so disastrously turned out to be a marvellous and significant year: I was back at Rolls-Royce again, working on

Hawker Horsleys being flown by 504 Squadron.

the most advanced aero engine in the world and continuing my passion for flying.

Hucknall Expands 2

It was stimulating to be back at work, especially to be included in such an enthusiastic team and I was grateful to Cyril Lovesey for remembering me. The task in front of us was to develop the 'R' engine to produce 2,300 bhp, an increase of 600 bhp on its performance in 1929. This was an ambitious programme and there wasn't much time or money to accomplish it. Money was still very tight after the depression; the government refused to fund the development and had it not been for the bounty of Lady Houston who put up £100,000 to pay for the project it would never have been able to go ahead. We were particularly keen to win the Trophy for if Britain won it again it would be the third time in succession and it would be won outright.

My first task was to help Bob Young in running a single-cylinder unit on the test bed. The cylinder was the same size as one of the 'R' engine ones but it was much cheaper to develop the cylinder this way than running a complete engine. The object was to extract as much power as possible from the cylinder by developing valve cooling, spark plugs valve and ignition timing and so on. We developed it to give 200 bhp and it made the most frightful din! The main engine incorporated the modifications resulting from the cylinder work, which was then tested for reliability. After we had finished the work on the single cylinder, I moved on to the main test beds to assist with the running of the engines. As time passed, the emphasis was put on getting the main engine to run for an hour at full power. It took many weeks to achieve this and many catastrophic failures took place. On one occasion the engine blew up at fifty-eight minutes due to a crankshaft failing and on another a con rod came through the crank case. It was all very exciting, and eventually a successful type test was accomplished.

The result after many long hours and much hard work by the team was that the programme was completed on time, the race was won and the world's air speed record was handsomely raised to 407.5 mph. The final bhp achieved was 2,600. A lot of the credit for this must go to Rod Banks who was the fuel expert working for Ethyll Export Corporation who concocted the fuel mix of 70 per cent benzol, 20 per cent petrol, 10 per cent methanol, plus 4 cc of tetra-ethyl lead gave the power required.

The Schneider Trophy story, quite apart from the glamour and excitement, the feats of engineering and the rapid development of both engine and airframe

The Supermarine S.6B having the 'R' engine installed.

has poignant lessons for any government. It is incredible to realise today just how significant that engine development work was and what might have been at stake had the lack of foresight and parsimony of the then government gone unchecked. If the generosity and enthusiasm of Lady Houston had not come to the rescue, the German fighters would certainly have had air superiority nine years later when the war started, and perhaps the Merlin which was a development of the 'R' engine might not have been produced at all! Memories are short, however, and there is a strong similarity in the situation today: the general economic slump has meant that there is less money available for research and development of aircraft and engines and that the numbers of combat aircraft are reduced—all due to Socialist government policy which refuses to face up to the threat from the USSR! And this time of course, costs are so high that no groups, let alone individuals, can bear the cost of modern development.

Once the Schneider Trophy had been won outright, the work on the 'R' engine ceased except for the preparation of engines for Captain George Eyston's record breaking car, Thunderbolt and Sir Henry Segrave's boat *Miss England* (George had bought the engines from the 1931 Schneider Trophy winner and these were being installed in Thunderbolt.) Development activity was centred on the Kestrel and Buzzard engines which had firm commitments overseas and with the RAF. The flight development was being carried out at the Royal Aircraft Establishment (RAE) at Farnborough, at the Royal Aircraft and Armament Establishment (A & AE) at Martlesham Heath and to a lesser degree at Tollerton, where Lovesey had obtained from the Ministry a Fairey 111-F fitted with a Kestrel and a Hawker Horsley fitted with a Buzzard engine.

They were flown by Captain Ronnie Shepherd, Chief Flying Instructor of the Nottingham Flying Club. The very first flights of the 'F' engine, the forerunner of the Kestrel, were carried out at De Havillands at Stag Lane aerodrome in a DH-9A by Hubert Broad who had some years before been in the Schneider Trophy team.

DH-9A; the very first flights of the 'F' engine, were carried out in this plane.

I was sent to the RAE engine flight at Farnborough for a watching brief on the work being carried out there on behalf of Rolls-Royce, who as yet did not have adequate flight test facilities of their own. This assignment included visiting the A & AE at Martlesham Heath where prototype aircraft were being tested, many of them using Rolls-Royce engines. Close contact was also being kept with Hawkers, Vickers and Faireys, who had experimental aircraft flying also using Rolls-Royce engines. This job was very interesting as it brought one into close contact with competitors and their products. There was keen rivalry at this time as there were many firms and the orders were few.

One of the most exciting aspects of this job for me was that it brought me into contact with the test pilots: George Bulman, Jerry Sayer, Chris Staniland, Philip Lucas who were my heroes. They were flying the Furies and Fireflies, etc. which were some of the most up-to-date aircraft in the world. My own flying was progressing steadily but I was sadly lacking in experience and flying hours. To try and get over this problem, I bought a De Havilland DH-53 from the RAE Flying Club. This little aeroplane, known as the Humming Bird was a single-seat, low-wing, monoplane, fitted with a Bristol Cherub two-cylinder engine of 35 hp. Its top speed was 82 mph, it cruised at 68 mph and landed at 38 mph.

Chris Staniland in the Fairey Firefly at Hendon.

One of my most eventful flights in it was probably the time when I flew from Farnborough to Hucknall via Brooklands and Sywell. There was a very strong head wind that made the going very slow and the compass revolved in phase with the vibrations of the engine which did not assist my navigation. Fuel was getting low and as I flew over Northampton it ran out and the propeller stopped right over the centre of the town! I managed to land in a field and after hitch hiking to a garage, I was able to fill up and complete my journey to Sywell where I was staying for the night. My journey to Newcastle was uneventful, and I parked the plane safely beneath the wing of a Vickers Virginia which was picketed down and surrounded by a rope fence. The next morning I arose bright and early to take my father to see my new plane. To my great dismay I found that some cows, which were allowed to graze on the aerodrome, had knocked down the fence, horned the wing of the Virginia and had chewed the ailerons and rudder of my little aeroplane!

It was while working at Farnborough that I met the girl who was to become my wife. She was a 'computer' in the aerodynamics department; in those days computers were human, female and much more attractive than the IBM variety! I had lots of opposition to overcome both from the Army and the RAF but I think I was probably helped by my new pale blue MG-J2; I used the obvious ploy and taught her to drive which enabled me to get my arm around her. My job at Farnborough came to an end all too soon, and having become well integrated with the various flight development programmes in the London area, I was sorry to be recalled to Derby and of course I was loath to leave my fiance too. The project I was given, was to investigate the feasibility of an automatic engine control system for the new Goshawk engine which was being developed for the new Day and Night Fighter Competition. I found it tedious reading up patent specifications especially not knowing much about it, and luckily I remembered that there was a Penn Automatic Mixture Control on test at Farnborough and a BTH ignition timing device also on test, both on a Bristol engine. The obvious thing was to go to Filton and talk to them about it. George Chick was most helpful and introduced me to Mr Davy of SU Carburettors who had a fully automatic carburettor with him which he was trying to persuade Bristol to adopt. It controlled the mixture with altitude and also boost pressure.

This seemed to be the answer to me and so I returned to Derby and reported to Cyril Lovesey and Hives that I thought they should invite Davy to see them to examine his carburettor. When I suggested to Hives that I felt that the carburettor, coupled with the Rolls-Royce boost control and the BTH ignition timer would give what was required, he remarked 'They won't let us get away with this!' They presumably were the design department. However, the result of the meeting was that they were impressed and an SU Carburettor was ordered. It was tested on a Kestrel and much to my delight it proved successful and was adopted. It was later fitted to all Merlin engines until it was superceded in 1943 by a Bendix which could operate under negative conditions.

My job having been successfully completed, albeit in an unorthodox manner, I was able to return to Farnborough. The urge to do this had been very strong and had undoubtedly had an effect on the solution. I was able to resume my courting and not long after I married my 'computer'.

The early days at Farnborough were great fun and although we reacted to our work responsibly there still seemed to be scope for various escapades. One I particularly remember happened at Brooklands when the F7-30 prototype fighter was due to do its first flight in the hands of George Bulman. I was there to carry out the engine checks prior to the flight and was with Ray Dorey, who was there as the technical expert on the steam cooling system. We both had interesting cars: Dorey had a Bugatti and I had an Invicta. A big end had gone on mine and as we had some time to spare until Bulman arrived we decided I should take it to Thomson and Taylors in Byfleet drop the sump and have a look at it before everyone arrived for the flight. I was underneath the car with the sump off when Dorey telephoned from Brooklands to say that Hives, Sydney Camm, Bulman and Lappin were on their way from Kingston to the aerodrome. Panic stations!

I had the sump back in minutes and headed for the hangar. I turned into the track just a hundred yards in front of them. I gained another two hundred yards inside the track and just got into the cockpit in time to start her up as they arrived! These sort of escapades may seem today to have been light hearted and irresponsible but they weren't in fact. The job was always done and one was always aware of one's responsibility to the other members of the team. Hives used to occasionally drop us sage words of advice such as: 'If there isn't much to do, don't hang around and make the job look untidy. Clear off and have a game of golf as long as we know where to get hold of you.' We never let him down and this is still very good advice today. Life was enjoyable and interesting; aviation was at an early stage and there was much scope for ingenuity and advancement. Hives encouraged both of these things.

Whilst doing my job I had been flying as often as possible to amass sufficient hours, and gain the different types of aircraft in my log book, necessary for approval to become a test pilot. I had flown on Avro 504s, Hawker Horsley Harts and the Humming Bird and eventually I felt that I had amassed enough experience and hours to apply for acceptance as a pilot. At our annual camp at Hawkinge it was arranged for the Air Officer Commanding of the Group, Air Commodore MacNeece-Foster, to fly as my passenger in a Horsley during a squadron formation so that he could recommend and further my application.

All went well until after landing in a cross wind the aircraft did a cart wheel. This was a fairly common occurrence as no brakes were fitted. My heart sank until the Adjutant Tubby Dawson came up and explained the situation. The AOC was kind and accepted it; the recommendation went through and I was duly approved by the Air Ministry to test the aircraft on loan to Rolls-Royce. This was a great and memorable occasion only marred by a remark made later by Hives who said that test pilots at Rolls-Royce were just flying testers and certainly not the 'knights of the air' they were considered to be by the aircraft companies. He had a special knack of bringing people down to earth and not letting them become too big headed. I became the first test pilot on the Rolls-Royce pay roll and at the princely sum of £4 10s. per week, was certainly no 'knight of the air!'

My new job was to fly the aircraft leased to Rolls-Royce which were then being operated at Tollerton by the Nottingham Flying Club and being flown by Captain Ronnie Shepherd, who was the Chief Instructor. The test programme was under the technical supervision of Cyril Lovesey, who was now in charge of the aero experimental department. Shortly after this Hives sent for me and told me that they had a Gnatsnapper coming up from Glosters. He wanted me to take it over and to get the steam cooling system working properly. He said, 'Take it as your project, fly it and do all the necessary work on it.' This meant that I had to get all the drawings out, study the aircraft and the engine installation and work out a programme of tests with Jimmy Ellor and Cyril Lovesey; quite the most interesting project I had had to tackle.

The Gnatsnapper was a single-seat fighter, a forerunner of the Gloster Gauntlet, and was fitted with a Kestrel 2-S engine, evaporatively cooled. This

The Gloster Gnatsnapper

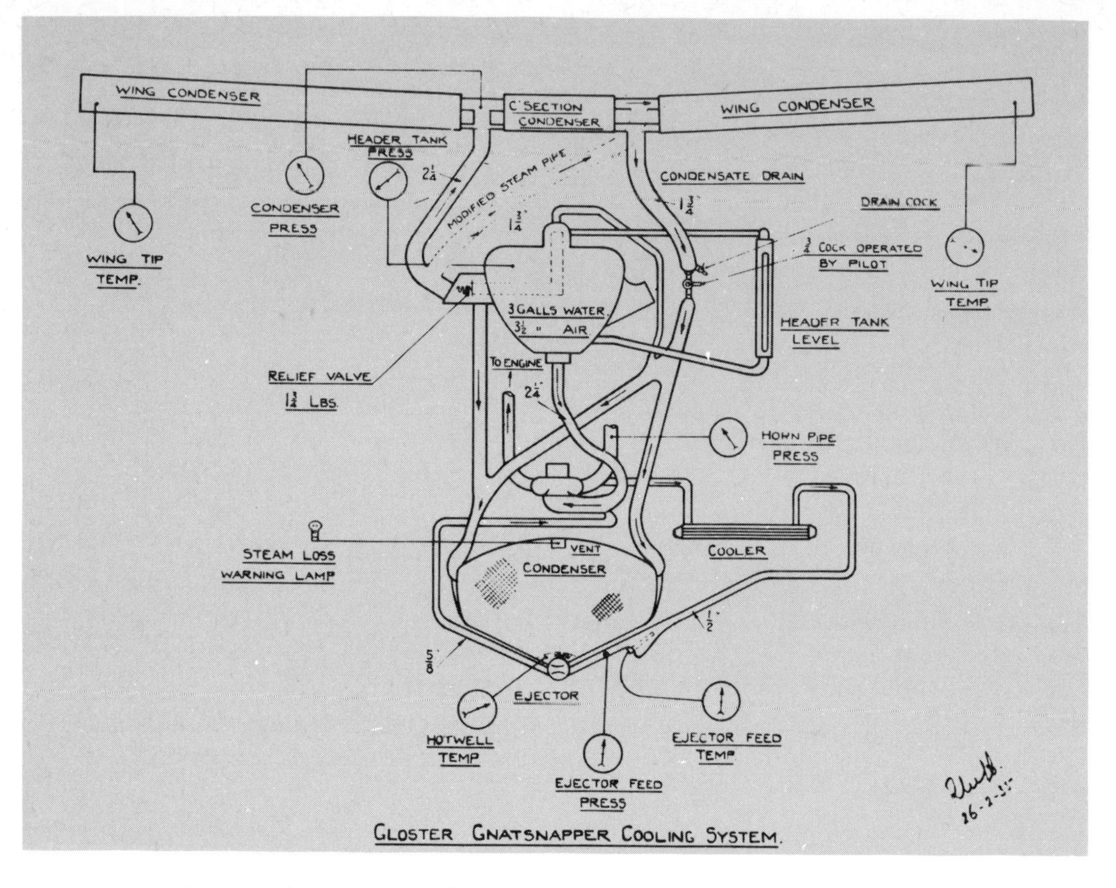

A diagram of the Gnatsnapper cooling system

system of cooling was sometimes referred to as 'steam cooling' and was being tried out and sponsored by Jimmy Ellor. It was achieved by leading edge wing condensers which were assisted on the climb by a small retractable honeycomb condenser; the idea was that this should avoid any cooling drag in level flight. Heat transfer from the hot spots in the cylinder head was improved by circulating the water at boiling point and then using the latent heat of evaporation to cool them (the loss of steam from the condenser resulting from bullet strikes would permit the engine to run longer than loss of water from a radiator), the system would also save weight as less water could be carried.

Considerable work was carried out on a number of installations, e.g. the Hawker Hart and Fury, the Gloster Gnatsnapper, Vickers four-engined bomber and their Short flying boat. The scheme was finally abandoned being superceded by using a glycol water mixture running under pressure at a temperature of around 110°C. This was a compromise allowing a smaller radiator than normal and when correctly cowled with an aerodynamic low drag cowling, the drag losses were considerably reduced. This was a simpler system eliminating the condensate pump and other complications.

There was no orthodox way of starting the engine on the Gnatsnapper: the way we did it was to get two people to pull over the propeller sucking in the mixture, then the pilot would wind a hand magneto in the cockpit with the switches on, hoping the engine would start! If it did not, the ground crew would be cajoled into swinging the propeller on contact. It was a laborious process and not without danger, particularly if the ground was wet or muddy. Something obviously had to be done about it and so I designed a starting handle to engage with an attachment on the engine which although made for this purpose had never been used by Glosters.

It was on 1 November 1934 that I sat in the cockpit with the engine running steady to take off. I remember reflecting for a moment on all the effort and concentration I had put into learning to fly, join the RAF Special Reserve and cajole the Ministry and Rolls-Royce to permit me to become a test pilot! My cup was full; I was thrilled even if a little apprehensive. It was a typical autumn day in the Midlands with ground haze caused by smoke and moisture which cleared rapidly as I climbed up through seven-tenths cloud to the blue sky above. The satisfaction of climbing up at a steep angle in this supercharged fighter with the engine running so smoothly and all the gauges reading as they should was most pleasing and exhilerating. I reached 20,000 feet in about twelve minutes; the lack of oxygen was a new experience as I had never been over 14,000 feet before. Then I noticed that the water pressure was fluctuating and the red warning light had come on showing that the condensate was building up and was not being pumped back into the engine. I had to get

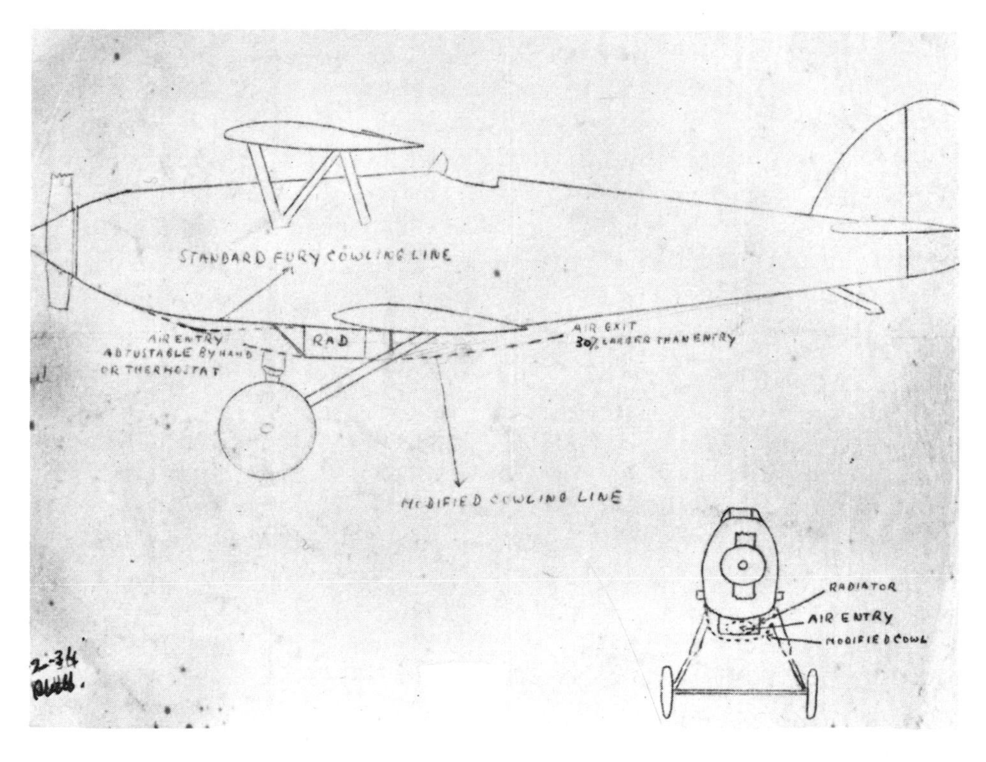

A drawing of the semi-dragless cowling for the Fury radiator in 1933

back to the ground soon as I knew that the engine would soon develop an internal water leak now that there was no circulation taking place, and would probably seize up.

I came down to the murk below the clouds with just seven gallons left. I began to feel anxious as the endurance was short and the ground haze was making the finding of the aerodrome difficult. Everything had been happening rather too quickly and I was beginning to wonder whether my test flying career was going to be terminated before it had begun when I found the aerodrome and managed to make a safe landing. Fortunately, although the engine was short of water it was undamaged. I resolved to be more careful the next time!

Being project pilot on this aircraft was a fascinating experience as all my knowledge was called upon to accomplish a satisfactory result. I had to fly the aircraft, diagnose the short comings of the steam-cooled system, design the modifications (with help) to the cooling system and get them made and fitted again and then fly the aeroplane. When the system finally functioned satisfactorily I wrote the report for the Air Ministry in order that the company could receive payment on the contract.

My ambition had been realised and the dream which had started at the Hendon Air Pageant in 1927 had come true. The important point which emerges here is that ambition can be realised, no matter how unlikely it may seem at the time provided one is sufficiently determined and the company is flexible enough to give encouragement. Rolls-Royce, with its broad outlook and enterprise, was an ideal institution in which an individual could develop. I feel that Hives was particularly responsible for this state of affairs as when he became more dominant, eventually taking over absolute control, his system produced many distinguished engineers whose ambition and ability had been encouraged. Thus a breed of hard working, dedicated and brilliant men emerged imbued with the traditions of Rolls-Royce and the human qualities of Hives. Herein lies the strength of Rolls-Royce and its greatest asset: people like Pearson Lovesey, Hooker, Lombard, Dorey, Hinkley, Huddie, Claud Birch, Morley and Ken Davies are just a few of the men who emerged with these principles.

Soon after the arrival of the Gnatsnapper, Hives decided to start an organised flight test department. A hangar was leased from the RAF at Hucknall aerodrome near Nottingham on the north-west side of the town. Offices were built and we moved our two aircraft from Tollerton. Ronnie Shepherd left the flying club and joined Rolls-Royce to become the Chief Test Pilot with me as his junior. Little was known about performance reduction and so we bought and studied an excellent book on the subject! Bill Horrocks, a mathematician, was sent to Martlesham for a few weeks to learn all about the methods adopted by the Air Ministry, and Harold Green and Harold Frearson, both experienced engine testers, were put in charge of engine servicing with Frank Parnell in charge of aircraft maintenance. From this nucleus a team was built up capable of handling the increasing numbers of aircraft which were sent to us for development.

Rolls-Royce Hucknall in 1935

In these early days at Hucknall, Cyril Lovesey was in charge. He had always been keen on aviation, owning a Moth G-AAEE, and he had long realised the importance of testing engines in the air as a complementary way of developing engines on the test beds. He felt it was all very well having an engine that ran well on the ground and completed type tests but that it was all a waste of time if it then gave trouble when installed in an aircraft. Until now the installation had been the responsibility of the different aircraft manufacturers, not all of whom produced a reliable system, with the result that most of the engine failures in flight were due to poor 'plumbing' of the cooling and oil systems. Hives felt that it was high time Rolls-Royce took over the design of the whole installation and was responsible for it.

An installation department was formed under Jimmy Ellor in 1934. He undertook to design all future aircraft installations based on data obtained from test apparatus which had been set up in Derby. His department now made it possible to experiment with such things as steam cooling which required quite elaborate test equipment which the aircraft companies had not been able to contend with. So that the installation department and the flight test department could between them, take the engines from the test beds, control the installation, and then submit them for flight test where the effects of atmospheric density, air temperature, aerobatic manoeuvres and drag reduction could be studied.

From this modest beginning a minor air factory soon emerged and over the following five or six years, a wind tunnel, a drawing and design office, a machine shop, comprehensive test rigs for all components, cold chambers, test beds, a propeller hangar, an electrical department and an administration department were all added to the original buildings. During the peak work period in 1942–3, some thirty test aircraft were on the establishment flown by ten test pilots and the original five people had grown to 1,640 people!

Jerry Sayer and the author in the Steam-Cooled Hart at Brooklands.

The PV-12 engined Hart

Our fleet of aircraft grew and soon we had three Hawker Harts, including the Steam-Cooled Hart K-1002, and a PV-12 Hart K-3036 (which was the test bed for the new PV-12 800 hp engine). This had a four-bladed propeller which coupled to the greatly increased power gave the aircraft a very high rate of climb and increased forward speed. Then came the High Speed Fury K-3586; this had leading edge condensers and a 650 hp Goshawk engine, and was the fastest aircraft in the country apart from the Schneider Trophy racers. It gave us much

Starting up a High-Speed Fury

satisfaction to fly these high-performance aircraft and we rarely lost an opportunity of demonstrating them whenever we got the chance. We also had a Gloster Gnatsnapper, an Audax and later two Merlin Horsleys.

We found that the drag of these biplanes was such, that hoped for improvements in drag reduction could not be measured and so it was decided to obtain one of the cleanest aircraft available so that we could note the results of such improvements as smaller radiators, rear facing exhausts, etc. The Heinkel 70 was chosen and a Kestrel engine was sent over to Rostock in Germany for installation and to replace the BMW engine. The Heinkel was a four-seater passenger and mail plane and was by far the most advanced and elegant aircraft available.

The story goes that before they installed the Kestrel into the Heinkel, they put it into their own Me 109 and possibly a Heinkel 113 fighter for their own edification! Herr Otto Cuno, the German test pilot, flew the Heinkel to Hucknall, accompanied by Shepherd. He stayed with my wife and I at our flat in Nottingham whilst he instructed us to fly our new 'toy'. We were all delighted with it as it was so much more advanced than anything else in England at that time. With the Kestrel installed, it would do about 300 mph which was 60 mph faster than the fastest fighters at that time.

Our experimental work when trying out such installation features as rear facing exhausts and ducted radiators with low drag cowls showed an immediate increase in speed. High temperature cooling, using either glycol or water under high pressure with smaller radiators, also showed improvements. So almost immediately, considerable benefits were apparent which when applied to our own British prototypes, the Battle, Spitfire and Hurricane, showed very considerable increase in performance.

Flying the Heinkel 70 fitted with the Kestrel engine.

 Ronnie Shepherd flying a Merlin Horsley.

While Cuno was staying with us he went to London to the German Embassy to record his vote for Hitler. We had many discussions about the possibility of war but it seemed very remote at that time. When war did finally break out he became a Luftwaffe test pilot and after the war when I met him again we had a most enjoyable evening exchanging tales of our competitive aircraft developments and discussing the various merits of the Me 109 and FW 190 against those of the Spitfire and Mustang.

Our next project involved the two Merlin Horsleys which came in for flight test. There was a bit of a panic on these as the new Merlin-engined aircraft were getting near to flight stage and it was urgent to get as much information as possible on the behaviour of the engine in flight. The three of us completed a hundred hours flying on this aircraft in six and a half days and this showed that the aircraft was reliable and fit for the new prototype aircraft.

Cyril Lovesey had by this time gone back to Derby to concentrate full time on the development of the Merlin engine. He had been replaced by Ray Dorey, who was a splendid organiser and really got Hucknall going as an efficient flight test establishment and factory.

The Hucknall test team in 1937; from the left are Purnell, Green, Shepherd, Heyworth, the author, Dorey, Trainer, Cowdray, A.I.D. Inspector, Fishlock.

The Build-up to War 3

Life was proceeding pleasantly with interesting test flying and with new types of aircraft including Fairey Battles appearing both at Hucknall and in the squadron where I flew at weekends. The squadron had changed from being a bomber squadron with Horsleys and Wallaces to a fighter squadron with Hawker Hinds and Hurricanes. We had our first Hurricane delivered for engine installation and development in March 1938. Meanwhile on the political front, the confrontation had taken place at Munich between Hitler and Neville Chamberlain, and this meant in our terms, that squadron training became serious: 504 Squadron, of which I was a member, was to be a front line fighter squadron.

The author flying a Westland Wallace in 504 Squadron.

The stable in 1938; from the left are the Fairey Battle, Merlin Horsley, Whitney Straight, Heinkel, H.S. Fury, Hart, PV-12 Hart.

The Hawker Hart, the Whitley and Hurricane

It was about this time, in May 1938, that Empire Air Day practice was taking place. I had been chosen to do aerobatics: anybody in the crowd who would pay sixpence could call me up on the radio and ask me to do an aerobatic manoeuvre of their choice. The rehearsals in the Hind went well, but, came the actual day, I had to fly a Hurricane as the squadron had by then been re-equipped.

Nevertheless, I felt confident, and on the way out to take off I was presented to Sir Kingsley-Wood, the Air Minister, who told me 'Always keep looking upwards my boy.'

'Always look upwards my boy!' From the left: W/C Hughes-Chamberlain, Sir Kingsley Wood, the author and Lord Sherwood.

Now there was an Air Ministry order out which said that no aerobatics could be performed below 5,000 feet because there had been several accidents in the past due to inexperienced pilots allowing too little height in which to pull out of a dive. When I came to obey the first request for an upward roll, I must have taken the Air Minister's advice too literally, for I zoomed across the aerodrome at nought feet, rose steeply upwards, performed the roll, came down again, zoomed across the aerodrome and repeated the action for some ten minutes or so.

On landing, I was in real trouble! I had contravened all regulations and I was told to report to the Air Officer Commanding at once. I did so and he turned out to be none other than Air Vice-Marshal Leigh-Mallory. Fortunately I had already asked my Commanding Officer, Squadron Leader Lord Sherwood (who was also Joint Under-Secretary for Air), if it would be all right to ignore the Air Ministry notice about the 5,000 feet level and he had replied 'Not to worry—after all we are the Auxiliary Air Force!' Unfortunately this did not stop us all getting into a lot of trouble—the Station Commander, my CO, the adjutant and myself. A summary of evidence was taken and went on every Saturday for several weekends. Squadron Leader Ronnie Lees, who commanded 92 Squadron at Church Fenton, flew down in his Spitfire each weekend to conduct it. I was later to meet him again under more favourable circumstances, and we are good friends to this day.

Leigh-Mallory eventually reported me to the Air Council for contravening regulations. Luckily, Lord Sherwood had me exonerated on the grounds that I was fully experienced and had been flying the Rolls-Royce Hurricanes for weeks. It was useful having a CO who was a Minister and on the Air Council too! I am also glad to say that Leigh-Mallory bore no ill will for having been overridden, for during the war, I was to be a pilot directly under his command!

The Battle and the Hurricane aeroplanes were the first of many Merlin-engined aeroplanes to be produced. Rolls-Royce had done well with the Kestrel which had gained a good reputation in the RAF in the last generation of aircraft and had brought the company right back into the aeroplane business after a somewhat lean period after the 1914–18 war when Napier, Armstrong-Siddley and Bristol had enjoyed the larger share of the market. Experience with the Kestrel in the squadrons had enabled the company to build up a good service organisation and customer relationship under Platford and Lappin. The perpetual desire to please the operator was the underlying theme which had been company policy in the early days of Rolls and Royce and Claud Johnson with the motor car. Now that the Merlin was to sweep all before it, it was felt that the reputation of the company must be maintained and the RAF at all levels must be kept satisfied. Hives, Platford and Lappin were to concentrate on this.

Hives sent for me and told me that I must now work for the service department and ensure that the pilots operated their engines correctly. The idea he had in mind was to extend the 'outside inspector' concept upon which the motor-car division had been operating for years. I was also to lecture to all squadrons on Merlin engine handling, to deal with their flying complaints and generally 'get alongside the RAF'.

I returned to Hucknall to discuss with Ray Dorey how I could fly for him and also visit the squadrons on behalf of Platford. He was fully co-operative and said that if I couldn't fly, I couldn't talk to the squadron pilots and so the sensible thing was to do both. Platford agreed and so the plan went ahead. Lectures were given to all the squadrons using Merlins and their aircraft were flown when they had complaints; the COs were entertained at Duffield House and were introduced to Hives, who always presided at dinner, and played snooker with them afterwards. On one occasion I remember Air Chief Marshal, Sir Neil Wheeler, then a Flight Lieutenant in a Battle squadron, asking Hives who a distinguished elderly gentelman was and what his position was in the company. He had pointed out Elliot and Hives replied, 'He is the Chief Engineer and he ensures the excellence of the design and perpetuates the doctrine of Sir Henry Royce.'

We also made a point of entertaining all squadron and flight commanders who were using Rolls-Royce engines and taking them round the works. It was important for them to see the engines being made and to appreciate the skill that went into the making of them so that they would not lightly mishandle them and cause unnecessary damage. These visits built up a close link between the pilots and engineers, and we took the opportunity of pointing out that their needs were of top priority and that their criticisms were welcomed. We wanted the Merlin to be developed to be as efficient as possible operationally. In the long term this close contact has been of great value. The squadron leaders of 1938 are now the Air Marshals of today and the bond of friendship has remained.

It was at this time that I ran into my first difficulties with administration, those rather unimaginative people behind the scenes who dictate who should work for whom irrespective of whether the job will be done satisfactorily or not. They can do so much harm by upsetting people and destroying initiative. Platford had brought in an assistant manager from another firm in order to cope with the expansion which was taking place. Hives, who had agreed to this new appointment, invited a number of us to dinner to explain the new set up. It transpired that I was always to report through him and not directly any more. I was most upset by this and went to see Hives the next day to remonstrate. He was very nice and with his usual charm said, 'Just try it—it won't be too bad—after all you don't have to sleep with him!' I agreed to give it a try.

It didn't last long: I had flown down to Northolt where the first Hurricane Squadron, No. 111 Squadron, was stationed, to see how Squadron Leader Harry Broadhurst was getting on having just taken over from John Gillan. On my return to Hucknall, the new service manager rang me to complain that I had not contacted his service engineer! An argument ensued. We were like oil and water, we just couldn't mix. We were taken before Hives who summed up the position immediately and rather than force me reluctantly to conform, he persuaded us to co-operate and suggested that in future we should report on service matters to Platford. This may seem trivial in retrospect, but it enabled us to do our jobs as we saw fit, without friction. This small matter of judgement was typical of the 'Boss' and the way he dealt with human problems.

The service manager, Percy Calvert, went on to do a grand job of organising an expanding service department; his service engineers became very closely integrated with the services all over the world and after the war, with civil airlines also. They were much relied upon for assistance and advice. A system of reprinting and storing vast amounts of technical service data, failures, complaints, faults and criticisms was set up. From this data bank the engineers at the factory were able to understand closely the operating conditions and how engines behaved under difficult climatic conditions etc. Once Calvert and I understood each other we co-operated well; he would call me in with my team of liaison test pilots to investigate flying complaints on the spot, anywhere in the world. We would fly aircraft and discuss the peculiarities with the pilots. It all worked very well and in the end the objective was achieved; a fine rapport was built up with the customer.

Once the Merlin engine was operating in the squadrons, certain problems arose which had not shown up very seriously either on the test bed or during flight test. The most serious of these was the internal coolant leak. The leakage occurred at the top joint between the cylinder flange and the aluminium head. It resulted in a serious loss of coolant and eventual piston seizure. It was thought to be partly due to uneven expansion of the cylinder head, caused by changes in coolant temperature during flight, such as high temperature on climb and low temperature during glide when throttled back. In order to maintain a constant temperature a thermostat was developed by Bill Martin-Hurst of the Teddington Control Company. We tested this at Hucknall, and Bill would sometimes fly with us to see how things were going. These thermostats were fitted on all Merlin-engined aircraft and although they did not cure the trouble completely, they certainly improved matters.

This defect was really brought forceably to our notice by 111 Squadron, who were the first fighter squadron to use the Merlin. Their mode of operation; climbing at full power to 30,000 feet then diving down, thus reducing the coolant temperature to the minimum, aggravated the condition which caused failure of the top joint. Fighter equipment was of the highest priority so special attention was being given to the squadron. The squadron was also in the news because John Gillan had flown his Hurricane from Edinburgh to London in forty-five minutes at a ground speed of over 400 mph. Admittedly he had a gale behind him, but the newspapers got hold of the story and I believe the Germans were quite worried about it!

Hives sent me to Paris to visit Hispano-Suiza to discuss their 'screwed in' cylinder liner, which he thought might be incorporated in the Merlin and also to take a thermostat with me to fit to one of the Fairey Battles. It was my first trip abroad for the company. I flew out in our Miles-Whitney Straight aeroplane stopping at Belgium where the Belgian Air Force was using Hispano-Suiza engines. I wanted to assess their reliability from a coolant leak point of view, before going on to Hispano in Paris. The reports were good and encouraging and so then I went on to Paris for detailed discussions on the possibility of a licence to incorporate the design into our engine. Prince Poniatowski, the Managing Director, and Louis Birkigt, the Chief Engineer, showed me round the factory and explained the details of the Hispano cylinder construction. I took the drawings and a scrap cylinder with me back to Derby in my aeroplane and handed them over to the experimental department.

The drawing office designed a new Merlin cylinder using the Hispano screwed-in cylinder liner; it was thoroughly tested on the bench. There were no coolant leaks, but at the higher operating temperatures and pressures on the Merlin, the heat could not be dissipated through the screwed joint to the coolant and it overheated and burnt, and so was discarded. The trouble with internal leaks was not completely eradicated until a new design of two-piece cylinder with a dry joint was incorporated later on during the war.

It was an interesting and exciting trip and a useful experience; I looked forward to further visits abroad. But at home, the war tension was beginning to mount, thus my next tour abroad was to be in uniform.

We in 504 Squadron spent the July of 1939 at our annual training camp at Duxford, taking part in a war exercise with our new Hurricanes. We were delighted with these aircraft, finding it easy to intercept the 'enemy bombers' which were being represented by Wellingtons, Hampdens and Blenheims from Bomber Command. By early August 1939, there was real tension in the air: the

504 Squadron at war station, Digby

newspapers and radio were gloomy about the European situation—Nazi Germany had just marched into Czechoslovakia and was now threatening to invade Poland. We wondered when Britain and France would demand a halt to this aggression, but we all felt that there really couldn't be a war. It was with a sense of relief that we thought back to Munich when Neville Chamberlain had gained the year's respite which had enabled aircraft production to expand. Before Munich, the RAF had no eight-gun fighters—all they had to rely on were Hawker Furies, Demons and Gauntlets and Gladiators. These all had two Vickers guns with the exception of the Gladiator which had four Brownings; they were no match for the Luftwaffe with their Me 109s. The year's grace had enabled the Hurricane to enter service in quantity so that now there were a dozen or so squadrons fully equipped with Hurricanes together with a lesser number with Spitfires. This was a much better state of affairs, for our job was to intercept Heinkel 111s and Dorniers, and a year before, these had been considerably faster than the British fighters.

We returned to Hucknall from Duxford in the middle of August to find that Rolls-Royce had become organised for a possible war. The hangars were camouflaged, the Rolls-Royce painted on the roof had been blacked out, and gun emplacements had been erected. A visit to the Derby works revealed that books of petrol coupons were being issued, anti-aircraft guns were being prepared around the factory, and balloon barrages were being tried out.

I was told by Hives that if war started and I had to go with my squadron, my pay as a flying officer would be made up by the company to equal my test pilots salary, and my job would be waiting for me on my return after the war! This was typical of the generous treatment one came to expect from the company. Knowing that my job was safe should I survive the war was very important to me, for in those days jobs were not easy to come by and I could still remember the days spent at home having been laid off when the slump was on. Peace of mind was important. Harvey Heyworth, the other Rolls-Royce test pilot was also called back into a fighter squadron. This meant that Shepherd was on his own until two more pilots could be found to help him with the expanding fleet of aircraft. Jock Bonnar and Wilfred Sutcliffe, both instructors at flying clubs answered the call and became test pilots in our places. Later Harvey was released back from his squadron to continue test flying at Hucknall where he continued, becoming chief test pilot on Shepherds retirement. His young brother Jim spent most of the war in Bomber Command where he was awarded the DFC and bar operating Merlin Wellingtons and Lancasters. He too was seconded to Rolls-Royce for test flying duties and after the war he eventually succeeded his brother as Chief. I first met Jim just before the war started when his brother brought him to the squadron mess for tea; he was on holiday from school! I suppose the aviation bug must have bitten him as it did the rest of us.

The squadron left Hucknall for its first war station at Digby on the morning of 26 August 1939. By the time we had landed at Digby, found our dispersal point, picketted down the aircraft and got ourselves organised in the Mess, some of our wives had arrived by car to have tea in the Ladies Room. Marjorie found a room at the local pub, and so we were ready for war.

That same evening, all pilots were called to attend a lecture by the Station Commander, Wing Commander Rogers. He told us that things were serious and that war was likely to start at any time. The Germans had marched into Poland and we and the French had issued an ultimatum which was likely to be ignored. A raid might come at any moment, and we were to be at half an hour's readiness. There were two blue lights mounted on top of the watch tower; if one was illuminated it meant that the enemy was half an hour away, and if they were both lit, a raid was imminent.

We all went to bed fairly early that evening, sleeping on camp beds in the commandeered married quarters. Suddenly the sirens began to wail! I leapt out of bed and went to the window to see what the blue lights were doing on the watch tower. There was only one lit up. This meant that we had to go to the air raid shelter as were not yet trained to be fully operational at night. Hurriedly dressing, I rushed off only to notice that there were now two blue lights showing, and that therefore there was probably a gas attack pending. By this time a number of pilots were assembled in the shelter. Suddenly Joe Royce, one of 'B' flight's pilots said 'I think I can smell gas'. We all agreed and put on our gas masks. Of course it turned out that there had been a mix up in the Group Headquarters Operations Room. They had mistaken a returning reconnaissance flight from one of our bomber stations on the radar for the real thing—but it was interesting that the effect of the propaganda we had been subjected to was so strong as to make us believe that we could smell gas when there wasn't any near at all.

Cases of mistaken identity were to become a frequent occurrence at this stage, when radar reporting and filtering was being developed. A week or two after the incident there was a big build up of fighters over the Thames estuary. A raid had been reported and a Spitfire squadron was ordered up to intercept. This was plotted as an extension of the raid and so another squadron was sent up. Again this was added to the original plot and so more squadrons were scrambled, and so on. Sadly to tell a Hurricane was mistaken by a Spitfire for an enemy aircraft and it was shot down into the Thames.

Victor Beamish receiving the DSO from His Majesty King George VI; Victor was the first person to spot the German cruisers escaping down the Channel from Brest.

Our CO, Squadron Leader Lord Sherwood, left us and went to take up his duty as Under-Secretary of State for Air in London. Our new CO was Squadron Leader Victor Beamish, the elder of the Beamish brothers, all of whom played rugger for Ireland and the Air Force. He had a reputation of being a hard task master and a ball of fire, and so we all felt rather apprehensive about his joining us, and felt that our amateur days were over. They were! He soon had us practicing air attacks, formation flying, night flying and air-firing. As a result we quickly became serious and dedicated fighter pilots and fully operational. All this was not without many amusing episodes, a few accidents and some reprimands to the less efficient. One of the things he insisted upon was the care of the aircraft. After each flight he made the pilots clean down their own aircraft, partly as an example to the ground crews, and partly to instill in us the importance of keeping the equipment in good working order. Our squadron soon gained the reputation for having the best maintained aircraft in the Group.

We were very fortunate to have such a splendid character and able officer to command us; it was a privilege to serve under him. We became devoted to him—I think we all realised that in him, we had somebody who could raise the squadron to a high pitch of efficiency in a short space of time, and this was important. Sadly he was later shot down over the Channel when he was on an offensive patrol.

While operating with the squadron I kept in close touch with the developments that were going on at Hucknall in the test flying world. I in turn was able to tell them of some of the operational problems we were having in the squadron—such as engine surging causing blow backs through the supercharger, the importance of quick starting, and exhaust flames and their effect on night flying. Dorey had taken on two extra test pilots to cope with the increase of work. Variable pitch propellers were now becoming available and were fitted in the squadrons. This coupled with increased boost gave a greatly improved performance to the fighters and we felt very confident that when the time came to go into combat, we would be superior to the enemy.

My time with the squadron was useful for another reason: I was able to get to know many of the other pilots in the Command. We were all flying officers or flight lieutenants at this time but as war progressed promotion was rapid and it wasn't long before many of one's friends became wing leaders, then station commanders and even group commanders. A year or two later when I was back with Rolls-Royce and keeping in close contact with the RAF I found it of the utmost value knowing the key people so well. I believe that mutual trust can only be built up in this way when people have known each other for a long time and in all sorts of places in the world. This close association with people also brought the Company closer to its customer and operator, and this was all part of the Rolls-Royce tradition.

In the early part of 1940 the squadron moved to Debden where there was another Hurricane squadron, 17 Squadron, and a Beaufighter squadron, 29 Squadron. We always felt rather sorry for the types in 29 as we felt there was no future at all in flying at night—we always used to say 'Only bloody fools and owls fly at night!' We would take it in turns with 17 Squadron to go foward to Wattisham to do forward readiness and convoy patrols. We spent most of the time waiting and playing poker in the watch hut. Often, when we did finally get sent off on a sortie, we would reach the convoy too late and the bombers would have left—they didn't hang about!

On one memorable occasion we were due to return to Debden when the weather began to deteriorate rather rapidly. Victor Beamish decided to lead us back in spite of the low cloud and rain and poor visibility. We set off, but as we

The Hurricane

proceeded the weather got worse and worse until we were literally just scraping the tree tops—too low for us to get radio reception and therefore a homing bearing. But Victor lead us on and more by luck than anything else, we found Debden just as it was getting dark. To this day I am surprised that some of the pilots did not get detached. I am sure it was because we were more frightened of Victor than the weather conditions!

There were two other outstanding pilots in the other two squadrons who were also the youngest on the station: Bird-Wilson of 17 Squadron and Bob Braham of 29 Squadron. The former, after totting up a considerable score of enemy aircraft became a renowned expert on the development of fighter aircraft, and the latter specialised in night fighting and was one of the top scorers—in the same class as John Cunningham and Desmond Hughes.

In the spring I received an order to return on loan to Rolls-Royce to carry on with my test flying and development of fighter aircraft. Having been on active service with the squadron, I remained in uniform, and at weekends I used to return to the squadron at Martlesham and continue at readiness on convoy patrols and scrambles after enemy reconnaissance aircraft. There was little activity and so after awhile when the squadron went to France I ceased to visit them and so continued my full-time flying at Hucknall.

My wife and I returned to our home at Blidworth near Nottingham only to find it flooded. We had let it out to local business man who had wanted to move out of Nottingham in case of bombing. There had been no bombing and so he had returned to his own home in the city leaving our house empty with the water turned on. The severe winter of 1940 froze everything up and we had arrived just as everything was thawing! It was in an awful state and the whole place needed redecorating—not easy in war time—and so we stayed at the Flying Horse Hotel until we could get the place ship-shape again.

The Rolls-Royce War Effort 4

On my first day back at Rolls-Royce I noticed that many interesting developments had taken place. Hucknall had expanded: we now had a fleet of aircraft consisting of every type of plane in service using Rolls-Royce engines, including some experimental ones which never went into service. We had several Spitfires, four or five Hurricanes, a number of Fairey Battles, a Merlin Wellington and a Whitley. The Battles made admirable test beds as they had had five hours endurance and so were used mostly for assessing engine reliability. The Whitley had been re-engined with a Merlin and was the longest range bomber in the RAF until the introduction of the four-engined bombers.

During the period I had been away with the squadron the number of pilots had increased. Shepherd was the chief and concentrated on flying the aircraft on engine development. I formed a team whose job it was to assist and also to visit the squadrons to test their aircraft after complaints and give them instruction on the latest techniques in engine handling. My team originally consisted of Jock Bonnar and Wilfrid Sutcliffe to be joined later by Peter Birch, Harry Bailey, Group Captain Stokes and Dick Peach.

We acquired two Hurricanes for flying to the various RAF stations we visited. These had 'souped-up' engines, the shape of things to come, which we allowed the pilots to fly, thus giving them a chance to know what was in the pipe line. Later on these were replaced by two Spitfires, a Seafire and a Mustang.

One of the major problems we had to deal with was the variation of fuel consumption between individual pilots; this could be as much as 20 per cent between the best and the worst pilots. Some bomber pilots had difficulty in reaching their targets and getting home with a sufficient margin of fuel for safety. The case of the fighter pilots was more serious because they usually flew in formation, the range of the squadron then being that of the worst pilot! Clearly there was an important task here to educate the pilots in the correct manner to operate their engines.

All operational aircraft by this time were fitted with constant speed propellers. This enabled the pilot to select both the optimum engine speed, and the throttle setting, to suit the flight condition. It also enabled them to do the very opposite! We found that to obtain the best fuel consumption under cruising conditions, the throttle should be wide open and the rpm pulled back

The flying liaison dept., 1944. From the left: R. Stokes, P. Birch, author, H. Bailey.

Left: W. Suttcliffe, right: Jock Bonnar

to give the required power. Thus 'low rpm and high boost' became the slogan. We found a number of pilots were doing the opposite because they thought the engine ran more smoothly that way!

At this point I was still joining my old squadron at weekends. The war at this time was confined, with the exception of several daylight raids on German naval bases, to leaflet raids over Germany by the RAF Bomber Command. The Germans had carried out reconnaissance flights over Britain and had made bombing and machine-gun attacks on coastal shipping, fishing vessels and light ships.

Then suddenly, in April 1940, the German offensive and the serious war started. The Germans invaded Norway and Denmark and then attacked Belgium, Holland and France. The RAF now took an active part in the war, coming face to face with the German Air Force at the evacuation of the British Expeditionary Force at Dunkirk. Shortly after this, in May, the Germans began an assault on the shipping in the Channel and began bombing the docks, aerodromes, factories and London itself. This developed into the Battle of Britain.

For us, it became a question of how Rolls-Royce could best help. All our test aircraft were fully armed and we were linked into the 12 Group operations room so that we could defend the Nottingham area. We chased aircraft on several occasions but never made a kill. It was more important to repair damaged fighters quickly and to get them back into squadron service. To do this, the many damaged aircraft were quickly collected and were either repaired at Hucknall or they were cannibalised to alleviate the spare part situation. In my liaison work, I visited squadrons finding out what parts were in short supply and were causing aircraft to be grounded and then hunted for them in our own stores. If my search was successful, I flew them straight back to the squadron and so got the aircraft serviceable again. During the crucial Battle of Britain, this was more important than carrying out routine test work and so this and the flight development work was resumed only later when the Battle was won.

There were lessons to be learnt from the performance of our competitors. The Hurricanes proved very effective against the German bombers and they were admirably backed up by the Spitfires which concentrated on the enemy fighters at top cover. But the pilots needed, and were demanding increased power. After the Battle of Britain, Lovesey and his team, backed by Elliot's designers concentrated on providing it. This was where the earlier development efforts of the Schneider Trophy, and later the High Speed Spitfire Merlin development paid off: the engine had a wide margin of strength left in the design as it had already run at over 2,000 bhp and key parts had been strengthened to cope with this. This reserve of power was soon to be tapped, and it was eventually to keep the RAF fighters ahead of those in the Luftwaffe until the end of the war. Looking back, it is interesting to contemplate just how different the result of the war might have been had it not been for the experience gained from the Schneider Trophy races from which the Merlin engines and the Spitfire evolved.

One of Hucknall's main activities now, apart from engine development, was the conversion of a number of Beaufighters to use a Merlin power plant. Also Mark I Hurricanes were being converted to Mark IIs with an improved Merlin XX with a two-speed supercharger. Again, later on a number of Spitfires

The Merlin Beaufighter

were converted to have Merlin 45s and later again to have the Merlin 61. This meant that Hucknall had grown into a small aircraft factory. It was therefore able to make a major contribution to keeping up the replacement of damaged aircraft during the battle.

Early on in the war Hives had started a Monday afternoon conference which consisted of all the key people concerned with the design, development and manufacture of the Merlin engine in particular, and any other engineering projects of importance. The main object of the meeting was to have a weekly review of progress and to highlight complaints from the services, to understand the extent of the enemy competition and to ensure adequate production output. The meeting was chaired by Hives and attended by the Managing Director, Sir Arthur Sidgreaves, A. G. Elliot, the Chief Engineer, Harry Swift, the Production Manager, and Bill Lappin, who was there to reflect the Air Staff views. Cyril Lovesey, who was in charge of engine development, was a key speaker at the meeting as he had to provide most of the answers on engineering progress. Ray Dorey represented the flight test and installation at Hucknall, Fred Hinkley spoke up for engine design, and Rubbra, his senior, was there as the watch dog for quality engineering, a job that he and Elliot had inherited on the death of Sir Henry Royce. Calvert represented the service department which played an important part in keeping the engines going in the squadrons and helping the RAF mechanics to play their part. I was invited to attend to give the views of the pilots and to enable me to keep the squadrons in touch with current developments.

They were good meetings and fairly informal. People were always referred to by abbreviations of their name so that Hives became 'Hs', Elliot became 'E', Sir Arthur Sidgreaves became 'Sg' and even I, a junior, became 'Hkr', etc. This enabled junior executives to speak in a candid way to their superiors without sounding disrespectful and without incurring the formality of Sir. It worked very well and encouraged people to speak quite freely. Attending these meetings was a great experience for a youngster. Critical decisions were being made all the time and there was a strong sense of urgency about all the business under discussion.

I recall one or two rather amusing incidents which took place. Once Witold Challier was asked by Hives to prepare a performance analysis by the following day. Challier thought this to be impossible and said to Hives, 'Do you expect me to perform miracles?' To which Hives replied, 'Certainly, that's what we pay you for!' Challier retorted, 'You remember the last person who performed miracles? Well, I don't propose to end up like that!' Hives had the last word and pointed out 'He did them for nothing.'

On another occasion it was rather a warm afternoon and feeling tired after a morning's flying, I was just dozing off when I heard Hives say, 'Hkr, how is the ATD Unit going?' Now the ATD Unit was an automatic timing device on the Griffon engine, but having been caught unawares, I confused it with the AFDU and replied, 'I'm not sure, I haven't visited there recently!' This was a bad *faux pas*. Everyone roared with laughter at my discomforture and I resolved never to fall asleep again.

The Merlin was performing very well in service and was gaining a fine reputation for reliability and dependability, especially now that the flame traps in the induction system had been developed to overcome the original blower surge problem and work was seen to be progressing on the coolant leaks. As my good friend Geoffrey Tuttle, CO of 105 Fairey Battle Squadron said to me, 'The Merlin sometimes does the most extraordinary things but it always gets you home!'

During the Battle of Britain and for some time after there were a number of big end failures. sometimes resulting in an engine fire which caused forced landings and sometimes the abandonment of the aircraft. A big effort was put into an investigation into the possible causes. Cyril Lovesey, who was in charge of the engine development, had tried everything he could think of on the test beds, including overboosting and overspeeding the engine, but the trouble eluded him. It was at one of our Monday afternoon meetings that I was asked if I thought it could be due to some peculiar flight condition. I said that I thought it could be and that I thought I could reproduce the failure. In fact I was sure it must be so, and so I said I was willing to bet on it. Hives said 'Done' and so the bet was on.

I went away and began my investigation by visiting some of the squadrons which had been experiencing the trouble and discussing their operating methods with them. Having done this, I thought I had found the answer. When doing a slow roll, the engine probably cut out as the aircraft became inverted, thus allowing the propeller to go into fine pitch. When the aircraft completed the roll, the engine would come on again with the propeller still in fine pitch and this would cause the rpm to exceed the maximum permissable. The oil pressure would fall to zero under these conditions and so would take several seconds to recover. It seemed very likely that this would be the cause of the bearing failures and so I thought I was on to a safe bet. All I had to do now was to demonstrate it.

I took a Hurricane with a Merlin 45 engine on which I was doing a hundred hours endurance test and proceeded to carry out a large number of slow rolls. After doing a hundred of them (by which time I was getting quite good!), the engine seemed none the worse. I tried again with a reduced amount of oil in the tank to aggravate the condtion, but after fifty more slow rolls there was still no change. We reduced the amount of oil once more and still there was no sign of failure. The next test was to do some very steep dives which would cause the engine to over-rev to 3,600 rpm with a fluctuating oil pressure. This test was repeated with less and less oil in the tank, but again there was no failure!—in all I did 167 slow rolls and twelve vertical dives under exceedingly severe conditions. After this the engine was taken out and examined and the bearings were found to be in good condition—I lost my bet!

Meanwhile a parallel test was being conducted at the Aircraft and Armament Experimental Establishment (A & AEE) at Boscombe Down and there, Wing Commander Boyd had devised a test which did fail the bearings, but which was hardly representative of service conditions. He went up to 30,000 feet, rolled on to his back, thus losing the oil pressure, over-reved to 3,600 rpm, and with a slight negative 'G' all the way down, held this condition. This meant that the engine was starved of oil and was exceeding maximum rpm for at least thirty seconds. That did it—and the connecting rods duly came through the side. Having thus confirmed the cause of the trouble to be air getting to the bearings instead of oil, suitable modifications were incorporated and so that particular problem was solved.

One of the many well-known pilots I came in contact with at this time was Douglas Bader. He was flying Spitfires and Hurricanes and I met him at Martlesham Heath where he was rather critical of certain shortcomings of the Mark 2 Hurricane, and again later at Tangmere when he had complaints to make about the behaviour of the Merlin in the Spitfire. He was of course quite right in his complaints and his forthright criticism helped me to impress upon the engineers back at Derby the importance of making the improvements.

By January 1941, with the Battle of Britain now over, the Fighter Command was beginning offensive sweeps over Northern France. Spitfires were mainly used for this operation as they were faster than the Hurricanes and more of a match for the Me 109s. It was soon apparent, however, that the performance of the Me 109s had improved—the 'F' model now being used was considerably faster and climbed quicker than the original 'E' model. The Focke-Wulf 190 (FW 190) was also beginning to make its appearance, with considerable success. Both of these aircraft exhibited certain qualities which were superior to the Spitfire and pilots were coming back from sorties reporting that they were being out-manoeuvred. It therefore became a matter of some urgency that the balance should be restored in accordance with the fighter pilots' criticisms. I felt that I and my team of test pilots should be able to help here. We were now

The author flying a Spitfire V.

in constant touch with the squadrons and were concentrating not only on the behaviour of our engines but also on how the aircraft were competing with the enemy in performance and tactics.

Help came firstly from two directions: the advent of the twenty-millimetre gun and the Gyro gun sight, coupled with improved aileron control, and the introduction of auxiliary drop tanks. These did much to even things out, but it remained up to Rolls-Royce to provide more power, with better carburation to back them up.

Lovesey and Hooker had worked hard on an improved supercharger, and when this was finally fitted to a Merlin and tested at Hucknall in a Spitfire, the results were most satisfactory, giving the aircraft a much improved altitude performance. This modified Merlin was called the Merlin 45 (I had used one of these engines in a Hurricane when testing the reasons for the big end failures), and Ray Dorey then took the project over and set up a line converting Spitfire 1s which then became known as Spitfire Vs. This increased power put the Spitfire back on top for awhile, but the Germans were not standing still and they were continuously improving their aircraft too. The Focke-Wulf 190 was just beginning to make its appearance which presented further problems as it had certain advantages over the Spitfire.

We had been very slow in overcoming the Merlin's inability to keep

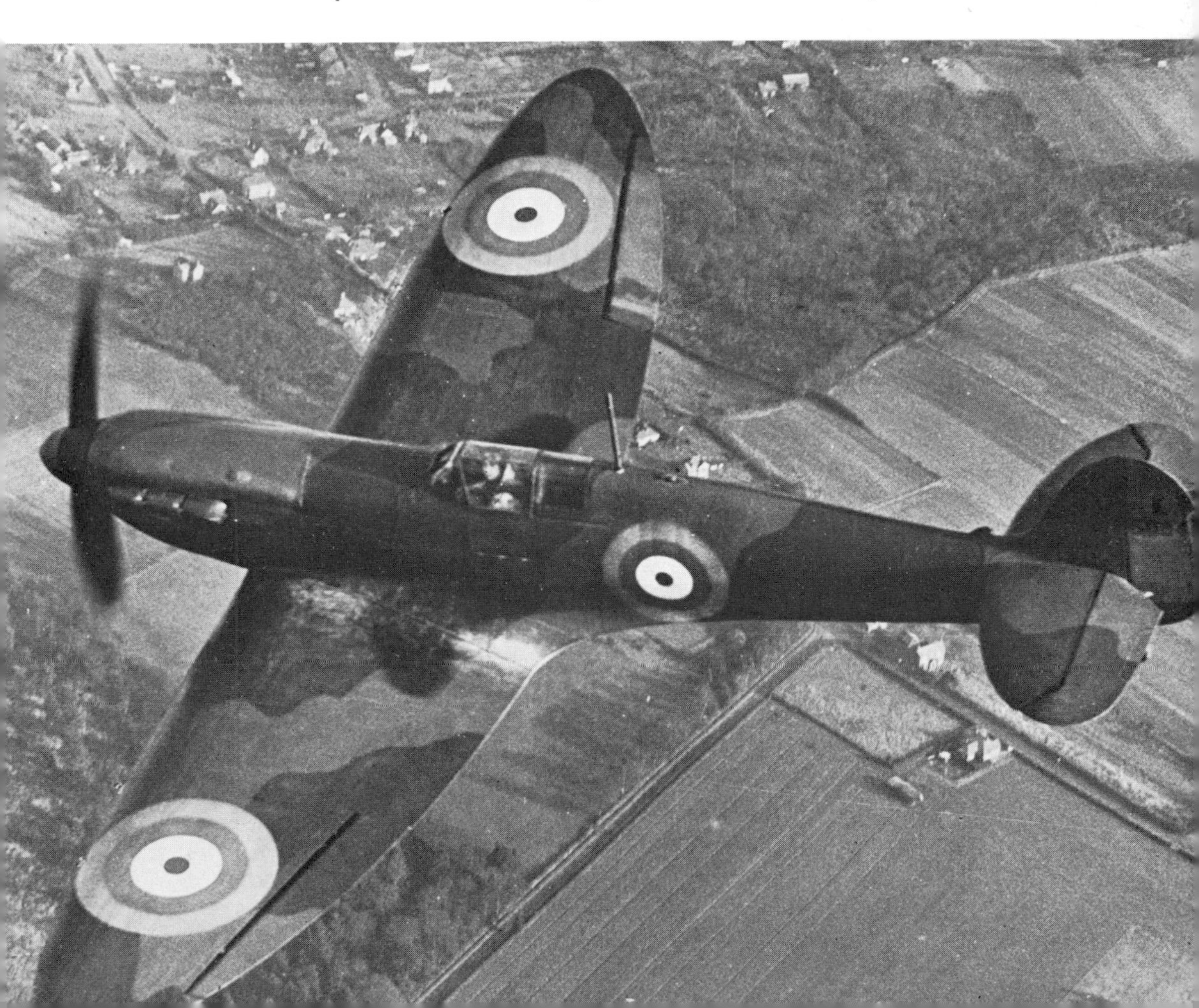

running under negative 'G' conditions. We had first reported it when flying the High Speed Fury in 1937, and it was not until Squadron Leader James Leatheart flew up to see us at Hucknall straight from a sortie to tell us exactly what he thought about it that urgent action was taken to try and overcome the fault. The Germans were working with the advantage that their aero engines had fuel injection. This enabled them to keep running under all conditions, so that when they were pursued they would push the stick forward and dive away without hesitation. Our chaps were at a disadvantage—the Merlin cut out for several seconds and so they were left behind. The solution to the negative 'G' carburettor problem continued to elude us although we flew several experimental models, they were not reliable for service use. It was while testing the Hurricane fitted with this carburettor that one of our test pilots had the engine fade when coming into land. He crashed and was killed instantly. I landed back shortly after and was given the job of breaking the news to his widow. My wife accompanied me to do this and we shall never forget it. Once she realised what we were trying to convey to her, we could see the utter helplessness and inevitability dawn upon her; it was final and she would now have to cope and bring up her young family alone.

A partial cure to overcome the effect of negative 'G' on the functioning of the carburettor was achieved by Miss Schilling, a scientist from Farnborough. She suggested that a washer with a restricted orifice should be fitted in the main fuel line. A complete cure was only forthcoming when the Stromberg carburettor finally replaced the SU type. This came in on the Merlin 60 series which was the next step to gaining continued performance superiority over the enemy. Again this stemmed from the diligent work done by Hooker and Lovesey on improving the supercharger.

The Merlin 60 series engine was a big step forward: it embodied a two-speed, two-stage supercharger and gave a large increase in power throughout the whole range of altitude. When it was first installed in a modified Spitfire 9 at Hucknall, the result was most impressive. Both the rate of climb and the top speed at all heights was increased. A batch of Spitfires were modified by Ray Dorey's team at Hucknall, and then Supermarines took it on for future production. We delivered the Hucknall batch to Squadron Commander Duncan Smith, one of the high scoring fighter pilots at 64 Squadron, Hornchurch. One of his flight commanders was Flight Lieutenant Mike Donnett, a Belgian officer who escaped to this country when the Germans overran Belgium, returned after the war and finally retired as Chief of Staff of the Belgian Air Force and was made a Baron.

Group Captain Harry Broadhurst commanded the station with Wing Commander Eric Stapleton as his Wing Leader. The Spitfire 9s were immediately successful and in order to confuse the Germans they were mixed up with Spitfire Vs so that they did not know which was which or how many Spitfire 9s were in the formation. The superiority now moved to the RAF. The Spitfire with the Merlin 60 series engine was further developed and was generally accepted as the finest mark of Spitfire produced.

We worked closely with the Air Fighting Development Unit at Duxford, which was commanded by Ian Campbell-Orde. I knew both he and his 'number two', Squadron Leader Ted Smith, from before the war when we were all members of the Auxiliary Air Force. The unit was equipped with all the latest fighters and its pilots were drawn from squadrons which had recently been in combat and so were fully conversant with enemy operations. Their primary job was to develop tactics to gain an advantage in air combat and to advise how fighters could be improved in handling and at which altitude they should be operated. I visited them frequently, exchanging views on engine development and what Rolls-Royce were doing to improve the Merlin. Whenever we flew an improved version of the engine at Hucknall, I would land at Duxford and ask them to fly it for their opinion and criticism.

I first became acquainted with AFDU when it was commanded by Wing Commander 'Tiny' Vasse when it was based at Northolt in 1937. It was only small then, and flew Hurricanes borrowed from 111 Squadron, which was the first to be equipped with this monoplane fighter. AFDU began evolving fighter tactics by flying in simulated combat with the new monoplane bombers: the Hampden, Wellington and Blenheim. These tactics then became the standard procedure for Fighter Command until the war got going and then they became drastically modified in the light of practical experience. As the war progressed this unit expanded and became the centre of gravity for fighter development. All new types came here for operational evaluation after being assessed at Boscombe Down for handling and performance. As the opinion of AFDU was sought on tactics and operational capability, there was naturally some rivalry and overlapping between the two establishments.

The Spitfire IX with the Merlin 61 fitted.

One instance of this was an incident over a modification on a Spitfire V. The rate of roll of the Focke-Wulf 190 was superior to the Spitfire and this gave the Germans a decided advantage. AFDU did some experiments with aileron gaps and removed the wing tips which helped considerably—but Boscombe Down did not altogether agree. This delayed the adoption of this modification in Fighter Command for some years, and it was only after it had been introduced in the Middle East in conjunction with some other engine modifications to improve low-altitude performance, and Air Marshal Leigh-Mallory saw how well the Spitfires were performing, that they were finally accepted in Fighter Command.

Of all the fighters that came to Duxford, the most renowned was the North American Mustang, which proved to be the optimum piston-engined fighter of all time.

It was 29 April 1942 when Ian Campbell-Orde (nicknamed 'The General') rang me to say that they had just received a Mustang AG-422 at AFDU and would I like to come down and fly it. He said it was a fine aeroplane and by far the best of the American ones he had tried. The next day I went down to Duxford and decided to make a day of it by taking my wife along, together with one of our technicians called Stowe who was an expert on fuel systems. We went by road for a change and arrived at about midday. We had lunch at the Red Lion Hotel at Whittlesford, a favourite hotel of ours where we had spent many happy times before the war when I had been on attachment at Duxford for the annual air exercises.

The General took me to the aeroplane and showed me all around it, explaining as he did so some of the history of how the specification had originated. As I flew the Mustang, I felt that it had a number of desirable features which the current fighters lacked. I was particularly impressed by its large fuel capacity of 269 gallons on internal tanks. This was three times as much as the Spitfire. I also liked the six .5 heavy machine-guns mounted close inboard in the wings, the light and effective aileron control which gave a high rate of roll and perhaps most important of all, its low drag which gave it a very noticeable increase in top speed over both British and German contemporary fighters.

With the low full throttle height of the Allison engine its overall performance was adequate for Army operation and reconnaissance duties at low altitude. However, one saw immediately the possibility of the Mustang as an air-superiority and long-range penetration fighter—if only it could be fitted with our latest two-speed, two-stage supercharged Merlin. If this was successful, it could be the answer to both the Me 109 and the Focke-Wulf 190, thus providing certain qualities that the Spitfire lacked. I discussed this proposal with The General and Ted Smith after my flight, and we agreed to put it up as a serious proposition.

On returning to Hucknall, I asked Witold Challier, our Polish performance expert, to estimate what the Mustang would do when fitted with a

The Allison Mustang

The Merlin Mustang

Merlin 61. He reported that there would be a greatly improved rate of climb and an increase of some 40 mph in top speed at 25,000 feet and above. This estimate, together with the fact that her tank capacity would give her longer range meant that the Mustang, when fitted with the Merlin, would be superior to any other fighter at that time.

One would have imagined there would be little difficulty in launching this project, but back at the Works there was considerable opposition. Perhaps it was the British characteristic of not readily accepting new ideas, or perhaps it was the 'not invented here' factor at work—one hesitates to say—but the unwillingness of people to accept new projects has always been a problem in industry, and it would seem that it still is today. My main difficulty in this case was to persuade the authorities to spend time and money to develop it. It was rather like trying to light a damp fire with a cigarette lighter; the initial flame is there, but the energy and effort required to dry the paper before it can catch fire is immense, but once alight is self-sustaining. And so it was with the Merlin Mustang—which as history shows became a conflagration!

I began my task by writing a note to Hives, with copies to the Senior Executives, outlining the advantages of the Mustang, and the estimated improvement in performance if the Merlin 61 engine was fitted. One of them circulated a note in reply to the effect that my proposal was unrealistic, that the Air Ministry would not approve and that there would not be any engines available anyhow as they were all required for the Spitfires— and why waste them on an untried, American built aeroplane!

However, Ray Dorey supported the proposal and agreed that I should speak to Hives to try and persuade him to authorise a trial installation. After I presented the case with all the force and conviction I could muster, he said, 'If you really believe that it will do as you say and be superior to the Focke-Wulf 190, then we must do it.' This was the go ahead. He spoke to Air Marshal Sir Wilfrid Freeman at the Ministrty and three aircraft were allocated to Hucknall to be converted to Merlins. The paper was now dry enough to burn well and so the system got going. Word was passed to North American Aviation (NAA), the makers of the Mustang, and they were instructed also to do a conversion. There was keen competition between Hucknall and NAA to be first to fly the Merlin version and bets were placed on it.

Robert Gruenhagen, in his very comprehensive book *The Mustang*, describes the earlier history of the plane as follows:

> 'The Mustang, which became the greatest piston-engined fighter ever built, began its career as a fighter nobody wanted. Commissioned in 1940 by the British, who were searching the free world for military aircraft, the U.S. Air Force took such a dim view of an aeroplane not even off the drawing boards that it tried to get North American Aviation, with its superior manufacturing capabilities, to build the Curtiss-Wright R40 which already existed. But NAA stuck to its guns, and in the unbelievable time of 65 days, designed and built the airframe of what was considered the most aerodynamically perfect fighter of World War II.
>
> 'The prototype Mustang first flew on October 26th, 1940, and on its fifth test flight it crashed and lay upside down with its back broken in a plowed field just short of the runway of the Los Angeles

Airport. Again the decision had to be made whether to continue building an untested air plane or to start immediate production on the P-40. NAA decided in favor of the Mustang.'

He also describes the conception of the Merlin-Mustang idea:

> 'The Royal Air Force interest in the Mustang-Merlin combination was initiated on April 30th, 1942, when the Commander of the Air Fighting Development Unit at Duxford asked a Rolls-Royce factory pilot, R. W. Harker, to fly a Mustang serial number AG-422, for evaluation purposes. Harker, in reply to questioning after the flight, stated that the only improvement that could be made to the air plane would be the installation of a Merlin. The British, desperate for any fighter as quickly as they could get it, were amenable to the switch.'

The competition was won by Rolls-Royce Hucknall by six weeks. Subsequent flight trials confirmed the Challier performance figures and so we now had a really potent long-range fighter which was superior to all other fighters. Now that it had been demonstrated at Hucknall just how successful a conversion it was, the RAF decided to increase its order and the American authorities were told all about it. They too decided to go into full-scale production for themselves and they adopted the aeroplane into the USAF.

USAF Mustangs

An interesting account of the role played by the Merlin Mustang in the daylight bombing campaign, appears in *The World at War* by Mark Arnold-Foster:

'If they were to bomb by day over Europe, what the Americans needed was a long range fighter. They had reached the same point of frustration that the Luftwaffe had reached at the end of the Battle of Britain. The bombers were vulnerable in daylight unless escorted by fighters. But the fighters did not have the range to stay with the bombers. The Allies, unlike the Germans, found a solution. It was the North American P-51B Mustang fighter, powered by the Rolls-Royce Merlin engine. The Mustang was probably the most remarkable combat aeroplane produced during World War Two. It represented the ultimate development of the highly-stressed internal combustion piston-engined aeroplane. The Mustang had originally been ordered by the RAF in 1940 direct from the American manufacturers. When it was first delivered equipped

RAF Mustangs

with an American Allison engine its performance was disappointing. Rolls-Royce believed firmly that what it needed was more power. Equipped with a new engine the P-51 turned out to be the best single-seat fighter in the world. Its performance was considerably better than the Messerschmidt 109 or that of the Focke-Wulf 190. Above all it could be equipped with long range fuel tanks and could reach Berlin. For the first time, the American B-17s were able to count on an escort all the way out and all the way back. After their defeat at Schweinfurt the Americans ordered an immediate expansion of Mustang production. They were in action with the 8th Air Force by December. By February 1944 the 8th Air Force was back in business. The Deployment of the Mustang was a turning point not only because it possessed the hitherto undreamed of

> capacity to stay with the bombers all the way but also because the Americans used it more intelligently than the Germans had used their escort fighters during the Battle of Britain.'

One day, soon after the Merlin Mustang was in full production, I went to visit Wing Commander Robin Johnson at Gravesend. He had just received a batch of Mustangs straight from the depot at Burton wood; the aircraft were standing out on the field waiting for the weather to improve so they could go on a deep penetration flight into Germany. I had lunch with Robin and we were walking up and down on the tarmac chatting, when we heard a noise like a pistol shot. Shortly afterwards we heard the noise again—it came from the parked aircraft. We went over to one of them to see if there was anything untoward. Whilst doing this there was another report close by. We looked at the aircraft and noticed that the engine cowling seemed to be bent downwards a little. We unbuttoned the cowling, and surely enough the engine mounting was sagging. On further examination we found that two of the engine bearer bolts had snapped and the broken pieces were in the bottom of the cowling!

On examining the fractures the metal was crystalline and very brittle and it was discovered that they had been wrongly heat treated. It was remarkable that just the weight of the engine had snapped them off. It was most fortunate that none broke in flight and that they were all breaking at roughly the same time. All the Mustangs were grounded, and the bolts were changed before they were allowed to fly again. Had the planned operation taken place, many of the aeroplanes would have been lost before they got to Germany and the cause could have been a mystery for a long time.

Mustangs after the war did a number of record flights, the most notable of which were those in the hands of Captain Charlie Blair. He obtained a Mustang which he christened 'Excalibur' with the prime intention of flying over the North Pole. It was while Charlie was planning this flight that I met him in the London office in Conduit Street. He was the Senior Captain on Pan American Airways flying Stratocruisers across the Atlantic and he was preparing for these record flights in his spare time. He called in to see me having heard that I knew something about the operation of the Mustang, to see if I could give him any information on the handling of the engine for the best possible range. I showed him all the performance curves I had kept and so we discussed how best to operate the engine. A fortnight later he called to see me again, having flown back to the States on his Stratocruiser and then taken his Mustang up to Alaska on a fuel consumption flight to test out the effect of the engine setting we had discussed.

A few weeks later he took a weeks holiday from Pan Am and flew Excalibur over to London on 31 January 1951 in 7 hours 48 minutes, breaking the previous record for the transatlantic flight as he did so. He again called to see me and told me about the trip. It had been an exciting one, but nearly disastrous. When he been flying at 27,000 feet his oxygen pipe had become detached and he had nearly passed out. Fortunately he had realised what was happening in time and was able to connect the pipe up again before he lost consciousness. I advised him to fly to Hucknall to have the engine checked over before flying over the North Pole, and this he did. The last stage of his flight was successful. On 29 May 1951, he flew over the Pole in 10 hours 29 minutes, a distance of 3,300 miles. The following day he flew from Fairbanks in Alaska to New York in 9 hours 31 minutes, a distance of 3,450 miles. He was awarded the Harmon Trophy for these flights which is the highest honour for aviation and is presented by the President of the United States. Charlie is probably the most experienced pilot I have ever met.

15,582 Mustangs were built in total. Rolls-Royce, having made this valuable contribution to fighter development, went on to improve the Merlin still further, thus helping the allied fighters to stay ahead of their opponents. Finally when the enemy sent over the 'Doodle Bug' V1 flying bombs the boost pressure on the Merlin was increased to +25 lb which enabled the Spitfire, Mustang and Mosquito to chase, catch them and shoot them down. Many variants of the Mustangs were made. Perhaps two of the most interesting were the double Mustang with two fuselages joined together by a stub wing, and the turbo-prop version which used a Rolls-Royce Dart engine. The latter flew as a prototype only and did not go into production.

The special relationship which had grown up between the RAF and Rolls-Royce was now even stronger than before because pilots felt that whatever they asked for in the way of improved performance to enable them to maintain air superiority was given to them. They were also confident that production was able to keep pace with the losses incurred. Hives was continually in touch with Beaverbrook and the Service Chiefs and so was always fully aware of top policy, and they too knew what they could expect from the Firm. This close contact with our customers and people of influence proved its value time and time again.

When the Merlin Mustang had proved itself to be a successful fighter aeroplane, I approached Hives about a possible rise in pay. He asked me why I thought I merited one and I said because I had initiated the Merlin Mustang to which he replied 'That's what we pay you for!' However he did give me another pound a week shortly after and I was delighted.

You Bend 'Em, We Mend 'Em in The Middle East

5

The war in the Middle East was hitting the headlines during the summer of 1942; it ebbed and flowed as first General Wavell advanced from Egypt in to Cyrenaica capturing hordes of Italians on the advance to Tobruk and then General Rommel reversed the process and drove the British back to Egypt once more. Towards the end of the year the British were again preparing an offensive —this time under General Montgomery, with the firm intention of pushing the Axis right out of Africa. A large build up of military equipment was taking place in Cairo and the Delta area where all available aircraft were being assembled. Craft that were damaged and had been lying in the desert were being salvaged and repaired and new ones were being ferried out from England and America via West Africa. This time there was to be no mistake about who should have air superiority!

Air Vice Marshal Dawson was the engineer in charge of all repair and maintenance work at Headquarters Middle East Command and was responsible for providing the required number of fully operational aircraft. He was working under extremely difficult conditions; the war was often only fifty miles away, communication with Great Britain was uncertain making delivery of spare parts and supplies unreliable, distances were vast, resulting in a lot of time being lost in travel, and there were difficulties in using foreign labour. In order to try and ease the second problem at least, he signalled Hives to say that he would like someone sent out to give assistance on Rolls-Royce engine problems. Hives sent for me and told me to pack my bags and set off for Cairo forthwith.

I was thrilled with the idea and started to make the necessary arrangements at once. I embarked on a BOAC Catalina flying boat on 31 December 1942. There were only five other passengers: Sir Arthur Rucker, the Minister of State in Cairo, Mr Stevens, the First Secretary, Air Commodore Warburton, the Air Attache in Chunking, Wing Commander Sellars, an RAF doctor and myself. We were a mixed bunch and we all got on together very well—this was fortunate for the only way to get to Cairo was via West Africa and the trip was to take ten days!

The Catalina G-ASDA piloted by Captain James (now Director of Flight Operations at BEA) took us first to Foynes, then to Lisbon where we saw the

New Year in floating on the River Tagus whilst waiting for a launch to take us ashore and thence to Bathurst, Freetown, and Lagos, where we had to await the *Ensign Empyrean* which was to take us on to Kano, Fort Lamy, El Fasher and Khartoum. Here we changed boats for the last time and were taken on an Empire flying boat *Corsair* to Wadi Halfa and finally up the Nile to Cairo. This was quite a trip and we were all rather exhausted and sorry for James Warburton who had to go on to China! It was my first trip abroad and to be away from England with her war-time restrictions and winter weather was rather refreshing. My only regret was that I had to leave my wife and young daughter behind, but I enjoyed being able to buy them nice gifts in Cairo; things which could not be bought in war-time England.

Having found a room and settled in at the Carlton Hotel, I reported to Air Vice Marshal Dawson at the Headquarters, who briefed me on the general position in the Command, the objectives to be achieved and the reasons he had sent for me. For my own part, the main object of my visit was to obtain first-hand information on the behaviour of Rolls-Royce engines operating under desert conditions, with a view to recommending any improvements that might be considered necessary. Some of the rather drastic Dawson modifications were explained to me; they were in the nature of an insurance policy in case an acute shortage of certain critical spare parts arose through enemy interference to the supply lines. One, for example, was the fabrication of big-end bearing shells made of dura-aluminium taken from damaged propeller blades! As the emergency decreased, so the standard of repairs improved, and salvage schemes reverted to more orthodox practice. This showed that Dawson's organisation was flexible. Having seen how well this worked in Egypt one cannot help but compare it with the organisation seen later in North Africa which was an equally good example of an orthodox procedure carried out in a most unimaginative way, the result being far from efficient.

Within the first day or two it became apparent that our fighters were considered inferior to the contemporary enemy aircraft as they had been able to choose the altitude which suited them best. The Me 109G was superior in performance to the Spitfire 5 at high altitude and the Focke Wulf 190 below 12,000 feet. The first problem therefore was to improve the performance of the Spitfire at low altitude where it was expected the next battle would be fought. The Command engineers had been trying to increase the top speed by setting the propeller pitch to allow the engine to run at 3,200 rpm. This simply made matters worse as the power curve peaked at 3,000 rpm and the propeller efficiency fell off too! The thing to do was to drop the full throttle height to 6,000 feet where the maximum performance against the FW-190 was needed.

I despatched a signal to Stanley Hooker at Derby asking, how much should be cropped off the supercharger rotor to give + 18 lb boost at 3,000 rpm at 6,000 feet? He cabled back stating 3/4″. Dawson immediately authorised three engines to be modified. We also took the opportunity of clipping the wings by removing the tips as advocated by Campbell-Orde at AFDU—this improved the rate of roll and thus made the aircraft more manoeuvrable. A locally designed airfilter was also fitted and this proved to be an improvement over the standard one. In a few days the aircraft were ready for me to flight test. I carried out a series of level speedtests and established an increase of 22 mph at exactly + 18 lb boost at 6,000 feet! I was rather fortunate during the last flight as I intended to try the ailerons for effectiveness at the maximum diving speed but decided to complete the levels first. All went well until the last test when the engine blew up at full throttle and I had to force land in the desert. Whilst waiting for transport to arrive, I wandered around the aircraft checking it over to see if I could see what had possibly gone wrong and I noticed that the port aileron was hanging loose—one of the support brackets had fractured!

The results of the speed tests however were good: we had established an increase of 22 mph. The 'brass hats' were delighted, and it was decided that a whole squadron should be converted. This squadron went into action commanded by 'Widge' Gleed and repeated the successful surprise attack on the Germans which the Spitfire 9s had made at Hornchurch. This influenced Harry Broadhurst who shortly after, when he commanded the Desert Air Force, instructed a whole wing of three squadrons to be converted. We hit a snag later on however when Air Marshal Leigh-Mallery, who was commanding Fighter Command in UK, came out on a visit to 'Broady' and decided that on his return to England, his own Spitfire 5s should also be modified. The problem was that these modifications were not properly authorised. For having been done locally in the Middle East, they had not been put through the usual procedure of type test and air-worthiness test at the Royal Aircraft Establishment (RAE). Eventually, however, the Spitfire 5s, which had not been replaced by the two-stage Merlin Spitfire 9s, were modified which much improved their effectiveness.

There was another instance of a local modification in the Middle East being highly successful. This time it was just before Alamein—the Allies were building up their supplies in the Delta area before the offensive began. The Germans were carrying out high-altitude reconnaissance flights with their Junkers 86 Ps. These flew even higher than the Spitfire 9s which had been unable to intercept them on similar flights in Fighter Command at home. The Spitfire 5s in Egypt were hopelessly out-matched until Air Commodore Smylie at Aboukir started to put his mind to the problem.

He had on his staff Flight Lieutenant Beauchamp who, before the war, had been one of Cyril Lovesey's carburettor experts. He looked into the problem and made up some specially tapered needles for the SU carburettors. These gave better altitude compensation and therefore more power: the compression ratio was increased by lowering the cylinder blocks on the crank case this caused the valves to hit the pistons and so they were scalloped out to give the required clearance.

The engines now gave quite a lot more power, but as this was still not enough, two aircraft were to be used, one to have just two guns, the other to have only a radio. This enabled weight reduction to be achieved and so this, combined with the extra power from the modified carburettors enabled the two aircraft to be flown in formation and be vectored on to the enemy, to climb to his altitude and so engage. The Germans were taken completely by surprise; they were unarmed and therefore an easy target.

This successful operation put an end to reconnaissance flights over the Delta area and once again demonstrated how development, unhampered by red tape and given the impetus of meeting the urgent requirements of war, could be most effective. Decisions had to be made on the spot, using imagination, improvising when necessary and keeping in mind the fact that every available aircraft and engine must be serviceable at all costs.

About this time an Me 109G in serviceable condition had been captured and was being evaluated in Lydda, the airport for Jerusalem. Dawson asked me to go up there and fly it—he suggested I take Air Marshal Sir Hugh Pugh-Lloyd's Hurricane and that he would mention this to him. I took the aeroplane, with its broad pennant on the side, and off I went to Lydda. On arrival I was met by my very good friend Wing Commander McDonald, who was godfather to my younger daughter. This was a terrific reunion and so we made plans to celebrate. Fortunately we couldn't get the Me 109G to start until late in the evening, just as the sun was going down, and so there was no possibility of returning to Cairo that night.

I flew the aircraft that evening and was impressed by several things, notably the throttle response of its petrol injection BMW engine which worked equally well upside down (no wonder there were many instances of this advantage being demonstrated in combat) and its rate of climb which was superior to the standard Spitfire 5, particularly at altitude. However, it was not nearly so good as the modified Spitfires at low level.

A captured Me 109E

Macarka and I spent our evening together at the King David Hotel where a rather amusing incident took place. As we entered the cocktail bar, we could see a Guardee Officer in his splendid number one uniform talking to his attractive girl friend. A few minutes later in came a rather scruffily dressed flying officer in a sandy battledress who went up to the bar and asked for a slivovitz. Moving over to the smart couple he blew the alcohol from his mouth, lit it, and shot a flame between them! Macarka and I were immediately ready for trouble, but to our amazement the Guardee laughed and said, 'Jolly good, chaps! I shall try that!' and he went up to the barman and also asked for a slivovitz. The barman asked 'Would you like the cheap one or the expensive one?' 'The expensive one, of course, my man!'—and this is where he made his error because it doesn't vapourise nearly so well! All that happened was that a trickle ran down his tunic with a weak blue flame—amid the hoots of laughter of everyone in the bar.

It was this evening also, that Macarka and I had the idea of keeping tally on the movements of our friends by autographing the lift shaft of the King David Hotel. We stopped the lift thirteen seconds after pressing the button to ascend and then wrote our names and dates on the wall. I also did this at the Carlton Hotel when I got back to Cairo. Soon the word spread around and there was quite a collection of names of well-known fighter pilots. Unfortunately no trace now remains as the King David was blown up during the troubles after the war and the lift shaft at the Carlton had been white washed the last time I passed through!

I spent that night with Macarka at this flat in Jerusalem and bid him farewell the next day as I set off for Cairo in the Air Marshal's Hurricane. It was a ghastly trip, very bumpy, bad visibility due to a dust storm, and all made much worse by a monumental hangover! A tyre burst after landing at Heliopolis and feeling rather sorry for myself, I arrived at Headquarters only to find that I was in trouble and that everyone had been looking for me. Apparently the Air Marshal had arrived just ten minutes after I had flown off to Lydda, wanting to fly the Hurricane off to Palestine in the opposite direction to do an inspection of one of his stations! Explanations were eventually made and all became calm once more. It was simply that Dawson, having told me to take the Hurricane, had forgotten to ask his permission—a difficult situation at the time but one which when ever I meet Sir Hugh today he always refers to with good humour.

Having completed the tests on the Spitfire satisfactorily and flown several Curtiss Warhawks fitted with American-built Merlin engines concluding with the flight in the Me 109G, it was decided I should travel by car along the coast road to Benghazi to see the trail of wreckage left behind by the retreating Germans after their defeat at El Alamein. This was an intriguing experience and very encouraging to see the destruction wrought by the 8th Army and Desert Air Force on Rommel's retreating German Army. Hundreds of burnt out tanks, lorries and destroyed aircraft littered the road and desert all around. The edges of the road were still mined: flags and markers having been placed where it was safe to walk. My companion and I took three days to reach Benghazi from Cairo. We camped out each night under the stars. Having arrived at Benghazi I found several friends there who kindly gave me a bed for the night; Bing Cross, whom I had known since the old Hawkinge days in 1933, was commanding the station. He had received rapid promotion from Squadron Leader at the beginning of the war to Group Captain and had collected a DSO and a DFC in the desert campaign. He kindly lent me his Hurricane and parachute as I wanted to go up to Tripoli where Fred Rosier was in command and Ian Gleed the Wing Leader was operating some of the low altitude Spitfires we had converted back at Aboukir.

I flew up to Bir Dufan from where they were operating and spent three days there flying with them. It was interesting to see the enemy aerodromes all ploughed up to prevent us from landing; there was little enemy activity as they were very much in retreat. I met Air Marshal Sir Arthur Conningham at his Headquarters; he was in command of the Desert Air Force at this time, soon to be promoted, his replacement was to be Sir Harry Broadhurst. The Commander in Chief saw me in his caravan and gave me a brief run down on the general situation and praised the good work that the Merlin engines had been doing. They were in use in his Spitfires, Hurricanes, Warhawks and Halifax and Wellington bombers. He was glad to have the low altitude Spitfires which we had modified back in Egypt and confirmed their superiority over the FW-190s and Me 109s. He said the Middle East Air Force had complete domination in the air and so the advance could not be stopped; there was close co-operation between the RAF, the Army and the Navy and this was most important.

I flew back to Benghazi and was greeted by an irate Bing Cross; he had expected me back the same day and as I had his parachute he had been unable to do any flying. He said, 'Taking a man's parachute is like walking off with a horseman's saddle.' It is a joke with us both today when we meet and usually starts off many reminiscences over the years. Twenty-five years later when he was Commander in Chief Air Support Command he took me on the inaugural flight of the VC-10 to Singapore, where we again had an amusing incident; but more of this later in the book.

I flew back to Cairo in a Blenheim and reported to Dawson. He told me that he had been posted to North Africa to organise the engineering and repair work there—the war in the desert was virtually over as the Germans had retreated to Tunisia. He took me with him in his Liberator, and on arrival said that he had sent Hives a cable saying he wished me to stay on his staff instead of returning to England because he wanted me to visit all the repair and maintainance units before he carried out his reorganisation. I was not too pleased about this as I had by that time been out there three months and I felt it was time I went home!

It proved to be an interesting visit as it was completely different from the desert. Bing Cross had come across also and had been promoted to Air Commodore and put in command of 242 Group which was the Air Component

The Spitfire V

supporting the First Army. The weather had been very bad for sometime and all the airfields were waterlogged thus curtailing aerial activity. The Rolls-Royce engines were again highly praised and kept going in spite of inadequate maintainance and spare parts. It was much appreciated by both pilots and ground crew that R-R took such a keen interest in their problems and were clearly intending to act to overcome their complaints.

By this time I was beginning to feel rather unwell and keen to get back to UK; it turned out that I had contracted jaundice and was turning a bright yellow! By the good offices of Tommy Wisdom, a well-known racing driver whom I had known before the war at Brooklands and who was now in charge of Air Movements, got me on a plane to Gibralter. This was quite an eventful flight; it was a USAF Dakota with General Lee and Staff on board; we ran into very bad weather at Casablanca, there was no navigator on board so they asked me to navigate.

On arrival back home I became very ill and was in bed for six weeks; the one consolation was that before leaving for Egypt the Firm had taken out an expensive insurance policy for me, this was typical of their generosity and thoughtfulness; it included a clause for illness contracted abroad and so I was getting sixty pounds a week which was very acceptable!

Whilst recuperating, various people came to see me and kindly kept me up to date with the developments which had been going on. I was particularly interested in the progress of the cropped-blower Merlins in the clipped-wing Spitfires with which I had been involved so recently. The figure which I had obtained for speed and rate of climb had been reproduced at Hucknall and this mark of Spitfire was later put into production as the VB with the Merlin 45-M engine. These aircraft were used in Fighter Command as a stop gap until all squadrons were re-equipped with the Spitfire 8 and 9 (fitted with the two-speed, two-stage supercharger Merlin 60 series) and a number of them took part in the Normandy landings.

Bombers 6

I finally returned to work eight weeks after my arrival in England and having spent months concentrating on fighters I felt that it was time I became more involved with the Bomber Force. Sutty Sutcliffe, Jock Bonnar and Peter Birch had been visiting the Bomber Squadrons and keeping in touch with their requirements but now that Bomber Command had expanded and was mounting heavy nightly attacks on Germany there were operational problems to be solved and engine handling instruction to be improved.

Perhaps it is as well at this stage of the narrative to go back to the years before the war and trace the development of the bomber aircraft and their increasing use of Rolls-Royce engines, for by this time, in 1943, Rolls-Royce were producing more engines for the Bomber Force than for the fighters. In the early part of the war, most of the bombers were using Bristol engines. The main exception to this was the Whitley, which although it began with the Armstrong-Siddeley Tiger, was later converted to the Merlin just before the war started. The Merlin gave the plane greater range with a heavier bomb load; we had the first prototype at Hucknall for test as the power plants had been designed and built there. Charlie Turner-Hughes and Eric Greenwood, the two test pilots from Armstrong-Siddeley used to come over every day to Hucknall where we used to carry out test flights.

The first Merlin Whitley squadron to be equipped was 10 Squadron at Dishforth, commanded by Bill Staton. I used to visit the squadron often to talk to the pilots and discuss engine handling problems with them. Apart from Bill Staton, one of their most famous pilots was Willie Tait, who was to lead the enormously successful raid on the *Tirpitz*. Bill Staton himself had a very distinguished career as a fighter pilot receiving the MC, DFC and bar in the

The Merlin Whitley Bomber

First World War and again in World War II when he received the DSO and bar. He and I became good friends, both being keen fishermen. On one occasion we flew to Southampton together to go fishing at Testwood (Hives having kindly suggested that we do so) and had a wonderful day catching four salmon, two seatrout, three trout, a chub, and a pike. He retired from the Air Force as Air Vice Marshal.

The Merlins worked very well in the Whitley which was very gratifying as nearly all its experience had been in fighters. A Merlin version of the Wellington was soon to follow as was the new four-engined bomber, the Handley-Page Halifax. The Halifax, which had four Merlins was the first of the four-engined 'heavies' to go into service. It was a great improvement in range and bomb load and enabled Bomber Command to carry out damaging raids in the Ruhr and occupied Europe. Bomber Command operated mainly from bases in Yorkshire under AVM Rod Carr, Air Officer Commanding (AOC) Ten Group. Roddie Carr was well known in the Air Force for having flown a Hawker Horsley powered by a Rolls-Royce Condor engine to the Persian Gulf. Unfortunately he didn't make it and he came down in the sea after 3,000 miles due to a fault in the fuel system.

We used to visit the Halifax squadrons fairly frequently, partly to give talks on engine handling and partly to help overcome some of the operating problems. It was here in 35 Squadron that I first met Leonard Cheshire. He was later to become famous as the CO of 617 Squadron when he was awarded the Victoria Cross having already won three DSOs and a DFC. A number of aircraft were lost on night operations under obscure circumstances and not due to enemy action. The cause of these disappearances, after long investigation at Boscombe Down, Farnborough and at Hucknall was traced to the aircraft having to fly on the minimum cruise threshold due to a combination of over-loading and high-induced drag. Under this condition, and during necessary evasive manoeuvres the rudder became inbalanced and locked over. When this happened it could not be centralised and so the aircraft got into a spin and became out of control. The problem was overcome by reducing the drag and overcoming the rudder over-balance. The final development of this aircraft was the fitting of Bristol Hercules engines which gave more power than the Merlins and the position in which they were installed seemed to suit the Halifax rather better than the Merlin.

There was always keen competition between Rolls-Royce and Bristol to get their engines into the different types of aircraft. This competition was very beneficial to the Aircraft manufacturers, and of course to the RAF—as if it wasn't enough trying to compete with the enemy! This competition between the two companies went on all through the war and all the bombers had either Bristol or Rolls-Royce engines fitted into them at one time or another. It was an advantage to have both air-cooled radials and liquid-in-line engines to choose from and this resulted in the best result being achieved. One plane that was always better with the Merlin rather than the radial engine was the Lancaster. The third new bomber, the Avro-Manchester, together with the Halifax and

The Vulture Henley Test Bed

Short Sterling was designed to take the new Rolls-Royce Vulture which was a twenty-four cylinder engine having essentially two Kestrels in the form of an 'X'. We had tested this engine in a Hawker Henley and Hawker Tornado, but when it went into service in the Manchester in 5 Group there were still a number of problems yet to be solved. The engine suffered cooling circulation problems at first and then developed big-end failures due to a defect in the oil circulation system. A number of aircraft were lost on operation due to engine failure and this gave the aircraft a bad name although the aeroplane itself was well liked.

We had a Manchester in for investigation, and one day, when I happened to be visiting Ternhill aerodrome to lecture to a Hurricane squadron, I was out at dispersal talking to Squadron Leader Gerry Edge, CO of 605 Squadron, when we noticed a Manchester approaching the aerodrome on one engine with the other on fire. It was our aircraft and it was being flown by Reg Curlew. He came in on the approach but did not have enough power to reach the runway. He landed short in a field and the wing hit a tree and was torn off. The fuel tank was opened and the aircraft burst into flames. We rushed over to the plane only to find it a blazing inferno with ammunition exploding in all directions. The Flight Observer, young Broom, came staggering out with all his hair singed. Gerry Edge and Teddy Donaldson entered the fuselage by the rear door to see if it was possible to rescue the pilot. Gerry got as far as the cockpit but found him already dead on impact.

Soon after this it was decided to abandon production of the Vulture and to concentrate upon developing the Merlin for bomber use and replace the two Vultures with four Merlins. This I believe was one of Hives' most difficult and wise decisions. It must take great technical judgement and courage to know when to abandon a major project—particularly in war time. This decision was particularly welcomed in the production department where it was going to be much easier and more efficient to be able to produce just the one engine. The four Merlins in the Manchester was a highly successful conversion; thus was the Lancaster evolved.

The Lancasters were operated by 5 Group and were based mostly at Lincolnshire, Nottinghamshire and East Anglia. The Lancaster was the best heavy bomber of the war; it carried the heaviest load at the greatest height and the fastest speed. It reacted very well to engine development. I spent quite some time with the squadrons at Waddington, notably 44 Squadron, commanded by Babe Learoyd VC, prior to my visit to the Middle East. One naturally spent more time with the first squadron to be equipped with a new type of aircraft as it was essential that pilots operated the engines correctly in order to get the best service and performance. It was also useful to be able to overcome teething problems quickly and so prevent the equipment getting a bad name.

In order to keep up with the increased production of heavy bombers which were now coming off the line, and to meet the demands of the mounting bomber offensive, Rolls-Royce now opened up new factories in Crewe and Glasgow. Merlins were now also being made in the USA by Packards. The standards to which these engines were built were of the highest quality and as similar as could be, although there were some differences due to specification and the marques being different. It is interesting to note that the pilots in some squadrons became critical and in fact connoisseurs of our engines—being able to tell which factory their engine came from. One factory had a phase of producing engines with a shorter time between failure due to some error in manufacture technique. This soon became apparent to the aircrews who used to try and avoid using engines from this factory until the trouble was overcome. The same sort of thing happened with the airframes, and a Spitfire from Eastleigh was usually favoured to one from Castle Bromwich.

So that by 1943, the development of bombers was fairly well advanced and the first problem facing myself and my team of liaison pilots was how to tackle the pilots and to persuade them to achieve maximum economy of fuel consumption. This was a never ending job. It was easy enough to impart the necessary knowledge to the first squadron, but there were losses and new pilots had to be trained. New squadrons were forming at a rapid rate and so one had to try and get the handling drill incorporated into the curriculum at the operational training units. To this end, Rolls-Royce set up a special course for the fighter pilots at Derby run by Heinings and Maddox and assisted by a few selected operational pilots who were on rest. By 1942, selected pilots from the Bomber Squadron were also going through the school.

Group Captain Hamish Mahaddie tells an amusing story of how when he was a very experienced bomber pilot having done a tour of operations on Merlin Whitleys, he was invited to the works at Derby to talk to the workers and to look around the factory. He went to the test beds to see a Merlin running and a young tester asked him to set the throttles and rpm for the most economical cruise. Hamish did as he thought using high rpm. The tester showed him the flow meter reading and then set the controls for high boost and low rpm. The fuel consumption fell dramatically, much to Hamish's amazement and he was converted from then on! He shortly took over a Lancaster squadron engaged on long range raids to Italy. As a result of this demonstration he ran a sweepstake to see who could return from these raids with the most fuel. He never won, and this surprised him. It wasn't until a long time afterwards that he found out why. Each man in the crew used to put in half a crown for these sweeps. There were eight in a crew and eight aircraft on most operations, so there was eight pounds per raid to be won. The organiser of these sweeps was a keen business man and a Central European. He and the Corporal who dipped the tanks had a little fiddle going. The aircraft in which the mid-European officer flew always came back with the most fuel left and so kept winning the money! The Corporal was posted to another squadron and couldn't bear to leave without telling Hamish!

The job of my team of liaison pilots which now included both experienced bomber and fighter pilots was to keep in close touch with both the Rolls-Royce school and the Central Flying School, so that these establishments would be kept up to date with the findings from our latest flight tests. These tests were carried out either in aircraft that were on endurance test or in the three Hurricanes and Spitfires that had been allotted for this purpose—they kept us in good practice and enabled us to keep in very close touch with the squadrons and to learn first hand what operational problems they were experiencing. We also learnt the latest tactics of the enemy and so could advise the factory to take the appropriate steps on engine development to maintain performance superiority.

Once the tempo of bombing had been stepped up and Germany was under night attack as well as day attack, it became necessary to improve the accuracy of the bombing and so the Path Finder Force was evolved. This was the special Group 8 which was under the command of AVM Don Bennett. It was formed by selecting various well-trained squadrons whose job it was to find the targets and to mark them with flares so that the bomber force could bomb on the flares and so hit their target. If the Path Finder found that the main force of bombs were not hitting accurately enough, they would drop different colour flares in the true positions and so correct the error. Lancasters, Mosquitos and even Mustangs were used for target marking. The pilots became very skilled and on numerous occasions, squadrons from 8 Group were selected to perform special and hazardous missions where great accuracy was required. Don Bennett was a law unto himself and was always demanding better performance from his engines.

617 Squadron in 5 Group was a particularly famous squadron. It had been commanded by Guy Gibson VC, Leonard Cheshire VC and Willie Tate, and later Micky Martin who was so successful on the Dam Buster raid and a pioneer with Cheshire on low level target marking using Mosquitos and Mustangs. I remember flying over to Willie one day when he was planning the

Tirpitz raid. He was walking up and down on the tarmac when I arrived. When I asked him why he was so pre-occupied he answered that 'he had a Battleship on his mind'! We set up his engines to give +7 lb boost for continuous cruise so that he was able to carry the huge 12,000 lb earthquake bombs destined for the *Tirpitz*.

One day when Micky Martin and Willie Tait had been visiting us at Hucknall to discuss our latest plans for engine development, I took them to lunch at the Black Boy Hotel in Nottingham. They had a collection of six DSOs and five DFCs between them. The waitress who was serving us said 'Excuse me, but what are all those press studs (i.e. bars) you have on your medals!' I must say it was very sobering to realise how little these decorations for bravery and contribtion to winning the war meant to people outside the services.

At the beginning of the war, the Bomber Force, unlike the Figher Command who had been effective right from the start, were unable to be a serious power of destruction, owing to the poor bomb load, range and inability to defend themselves against enemy fighters. Daylight raids by Blenheims, Hampdens, Wellingtons and Whitleys had had to be discontinued due to the heavy losses incurred and were put on night operations. This reduced the losses but their bombing was inaccurate—only 3 per cent of the bombs fell within five miles of the targets. When the first heavy bombers, the Halifax, Sterling and Manchester had been introduced, greater loads had been carried but over 700 aircraft were lost by the end of 1941. The bombing effort had to be reduced as these losses could not be sustained. This was only increased again after Air Marshal Harris took over command of the Bomber Force and more bombers could be produced from the shadow factories and the Lancaster could be available in greater numbers. He then chose targets close to the coast line of Germany to assist navigation and bombing accuracy. In March 1942, 230 aircraft raided Rostock on the Baltic coast and a few nights later Lubek. These raids were highly successful, doing great damage and at the same time putting heart into the British people who at last felt they were retaliating.

The new navigation aid 'G' was now coming in and this enabled the crews to know their position in bad weather and to bomb through clouds. A 1,000 bomber raid was mounted on 30 May, the target was to be Cologne. This was a milestone on the way to stepping up the offensive with the intention of crippling German industry, and was the forerunner of many more mass raids. However the Luftwaffe had not been standing still: they had greatly improved their night fighter tactics and their anti-aircraft defences were also very heavy. Together these resulted in heavy bomber losses. Harris decided to carry out a very heavy raid on Hamburg using 'window' to counteract the enemy Radar. Window, now known as chaff, was a then secret device consisting of millions of tiny pieces of metalised paper which cluttered the enemy radar and so masked the signature of the aircraft. The result of this raid was astounding; so much heat was generated by the fires that a fire storm was caused by the air rushing in to feed the flames and this added greatly to the disaster causing

30,000 casualties. Albert Speer, the German Minister of construction and armaments, is reputed to have said that had this sort of raid been repeated six times Germany would have had to capitulate. Unfortunately, Bomber Command was not able to do this. We lost 1,000 bombers in four months in 1944 so the battle was not all one sided.

The Americans, who had been doing daylight precision raids while the RAF were operating mainly at night, were having very heavy losses to enemy fighters until the advent of the Merlin Mustang which was able to provide fighter escort as far as Berlin. Spitfires and Thunderbolts had been able to go with them so far, but lack of range prevented them from accompanying the bombers right to the farthest targets. The Merlin Mustang was able to go all the way and back again with the bombers and was able to master the German fighters causing them severe losses. The Mustang, having the range of the bombers and the performance of the best fighters, was the decisive factor in reducing the losses to sustainable proportions in daylight operations. The RAF also was able to operate in daylight and so bombing went on all round the clock; fires started at night were able to be stoked up during the day and still be alight for the next night. It was Marshal Goering who is reported to have said that once that bombers could be escorted over Berlin the end was in sight.

By the end of the war, when the bomber offensive was at its peak the Lancaster and the Mosquito were the main stays of the command. Both used Merlin engines although some Lancasters did have Bristol Hercules, their escorts were mainly Mustangs and for shorter sorties Spitfires. It was a remarkable achievement by the company to have built so many engines, and to have been chosen to power so many different types of aircraft especially considering we had started the war with only the Whitley bomber. To develop an engine to give so much power was also an achievement we were proud of: the Merlin gave 1,000 hp in the Whitley, and finished the war in the Mosquito giving 2,000 hp. The increase in bomb load and range due to this increase of power was very great.

We at Rolls-Royce felt proud to know that we had played a crucial part in enabling the Command to achieve success.

The Avro Lancaster with four Merlin engines

World Wide Liaison 7

Our work in this country was still involved in the never-ending battle to standardise the procedure of handling the engines for maximum efficiency under all conditions. In addition to the Engine Handling School at Derby, I had instituted an engine-handling panel. This consisted of Elliot, the development engineer of the particular engine under review, the Principal of the Rolls-Royce Engine Handling School and myself. Having decided on the optimum method of operating the engine in a particular aeroplane (a specimen of which we would have already flown at Hucknall), we would discuss the method with the RAF Handling Squadron commanded by Wing Commander Fryer at the Central Flying School, and the method would be adopted and written into the RAF manual. This was very important for if the engines were not handled correctly, some of the advantages of improved design and development could be nullified. By continually visiting the squadrons we were not only able to correct faults but to learn from operating experience how to improve the techniques.

One other thing we also did was to design a linkage which would combine the throttle with the rpm control so giving each aircraft the same fuel consumption. We had this fitted to a Hurricane at Hucknall for air test; it worked well giving a very good compromise and relationship of rpm to boost pressure. Full boost at full rpm was obtained, and for cruising, low rpm at full throttle, this gave very good fuel economy. I took the aeroplane round the squadrons and canvassed the opinions of operational and experimental pilots at Farnborough and Boscombe Down. At first the scheme was not at all popular. The experts all said they could do better with separated controls. We countered this by arguing that they might be able to when they could give their concentration to it all the time, but what about formation flying and looking for the enemy, and wouldn't it be an advantage for all the pilots in the formation to return home with the same amount of fuel in their tanks? It was never universally adopted in the RAF but the Royal Navy did have it in their Seafires.

We next turned our attention to the Royal Navy who now had a large number of Sea Hurricanes and Seafires operating on the Carriers. To do this effectively, we felt we should become attached to the Navy for a short while to get to know their own special problems. This naturally meant that we would have to do some deck landings!

Harry Bailey and I were duly posted up to HMS *Inskip* near Preston to commence a short deck landing course on Sea Hurricanes. The drill was to do a number of assimilated dummy deck landings (ADDLs). The technique was to approach the runway, which had the silhouette of an aircraft carrier marked on it, just above the stalling speed with the engine on and the nose up—the aeroplane then being in the landing attitude. On the runway stood the Deck Landing Control Officer whose job it was to signal to the pilot, by means of bats, whether or not he was coming in at the right attitude. When he was satisfied that this was so, he gave the signal to 'cut' which meant the pilot had to cut the throttle and that he should then be in a position to pick up a wire and land on the deck. The batsman was able to give signals whether to increase the angle and speed or vice versa as the case might be. It is at first rather strange to approach the runway so slowly, and to do what the batsman tells you. RAF trained pilots usually get accustomed to coming a little fast as there is no need to land short and it also provides you with a margin of safety. Naval pilots however were trained from the word go to adopt the deck landing technique. After having passed out at *Inskip* on ADDLs we flew up to Abbotsinch to rendezvous with the training carrier HMS *Ravager*.

There were three of us, Harry Bailey, myself and a squadron leader from Boscombe Down, who was CO of the Naval Flight. He had been telling us all about naval aircraft he had been testing and so we rather looked up to him! Harry Bailey seemed the most at ease and I was slightly apprehensive. The day arrived and we were despatched off to the vicinity of Ailsa Craig where we found HMS *Ravager* looking like a speck on the ocean. We received the signal to come and land on. I was in the lead and just as we were turning to come in on the approach, we got the red light to indicate that the ship had changed direction. When the manoeuvre was completed we received a 'green' and Bailey landed on first. Our squadron leader friend went on next and then it was my turn.

The sight of the propeller wash behind the ship doing twenty knots was quite awe inspiring and so I edged on a few extra knots of my own! The batsman gave me the go slower sign and then the 'cut', but alas it was too late and I floated over the arrester wires, hit the first barrier which broke my propeller and pulled the undercarriage back, and went on through the next barrier and into the tailplane of the Sea Hurricane. This had landed just before me and my plane pushed it (and the pilot, who was just getting out), on to its nose and to within six feet of the front of the flight deck! Harry Bailey's aircraft had been taken down the lift just in time. This was the first ever crash I had had and I must say that I was rather shaken and surprised. The navy didn't seem to mind and the Commander, Johnny Levers, invited me down to the ward room for a pink gin. We would now all have to qualify on Harry's aeroplane as the other two were badly damaged.

After lunch Harry completed his six landings very neatly. Then the squadron leader took off, and he too must have seen the hazard for when he came in to land he was also too high and had to be sent round again. He did this

Through the barrier! The author on deck landing course in the Hawker Hurricane.

five times and the Commander flying told him that if he didn't make it on the sixth time he would be sent back to base. I kept my fingers crossed as this would mean that I wouldn't then have the opportunity of completing my landings. He managed it the sixth time and handed the aeroplane over to me saying 'Honour is satisfied.' I must say that as I was strapped in I felt like a condemned prisoner, but once I opened the throttle, left the deck and become airborne, my confidence returned. I came in to land and caught the first wire. It was a bit short, but I was on the deck. It is an extraordinary experience being alone, high up in the sky, seeing just a tiny speck in the ocean with a white wake behind it and then within minutes of hearing a voice in the earphones saying 'OK to land on' clanging on to the deck and being immediately surrounded by sailors helping you out of the aeroplane. From one environment to another in such a short space of time is truly impressive.

I did another four landings which were all right and I qualified with an average assessment. I was disappointed with my performance. I felt I hadn't done too well, especially as Harry Bailey had proved to be so good. At first I thought it might be tenseness due to the fact that I was awaiting news of the birth of my second child, but I later discovered that most firms' test pilots, when doing the same course had either hit the barrier or damaged the aircraft, and so I put it down to not being used to this rather peculiar technique. I always had great respect for Naval Aviators after this!

My team of test pilots concentrating on flying liaison duties now consisted of Peter Birch DFC and bar, who had had a distinguished career in Bomber Command, flying Hampdens, Manchesters and Lancasters, Harry Bailey, an ex-Rolls-Royce apprentice and fighter pilot, and Tony Martindale, an ex-Rolls-Royce experimental motor car tester, who had an AFC and bar for distinguished test flying at RAE Farnborough. He had the distinction of diving a Spitfire vertically from 30,000 feet and reached a Mach number of .92 when the reduction gear and propeller came off. He force landed in a field on

The liaison team: from the left are Bailey, Peach, Brigginshaw, author, Stokey, Martindale.

fire, managed to get out and then went back to rescue the baragraph! Group Captain Rendell Stokes DFC, who had recently retired from the RAF in the Middle East, was also in the team along with Dick Peach, an ex-Rolls-Royce apprentice, who had joined the navy and served in the Fleet Air Arm. Sutcliffe and Bonar had left us: the former to go back into routine test flying and the latter to go to Napiers as a test pilot.

We were now able to cover all theatres of operations: Europe, Africa, India and the Pacific. I will quote one example of a successful tour by Peter Birch which showed the importance of correct engine handling control and how it could greatly improve the operational efficiency of a squadron or type of aircraft. We had received a signal from the Navy in South East Asia Command requesting help on the operation of the Barracuda to obtain a better range. Peter Birch was selected to go out to Ceylon to investigate.

He had quite an eventful journey out in a RAF Dakota, taking fifty-eight flying hours to get there. The Captain of the aircraft was slipped every other landing so that Peter was a continuous second pilot! On arrival at Ratmalana in Ceylon he found a long row of Barracudas parked with their wheels embedded in sand and generally in poor condition, due to long exposure and lack of maintenance. He sent a cable back to us 'Miracle required not genius!' However, he had one fitted with a flowmeter so that he could carry out some fuel consumption tests and compile some range figures.

After experiencing some difficulty and embarrassment coping with naval aviation regulations, such as how 'not to approach the landing strip by flying over a Capital Ship' he managed to produce some figures. After discussing them with the Navy and recommending the best method of operating the Merlin engine at the optimum speed, he did a long-range flight over the sea to prove his findings. The result was much better than had ever been obtained, and so he was asked to lecture to the pilots on how the squadron could obtain the same results. This was all very satisfactory and again demonstrated that personal attention and instruction by the firm was indespensible.

An amusing incident followed this however! The Admiral was so pleased with Peter's results that he invited him to join him for dinner. The problem was that Peter was totally unprepared for anything of this sort and had nothing suitable to wear. People in the Navy are renowned for being immaculately dressed, and this occasion was no exception. I believe Peter's red sea rig consisted of his khaki battledress trousers, a cummerbund and a white shirt (borrowed)!

While Peter Birch was in Ceylon Harry Bailey was despatched to visit the Air Force in Burma to demonstrate engine handling on Spitfires, and I went out to North Africa and Italy to keep the pilots up to date on the latest techniques of handling the latest two-stage supercharged engines. There had been some difficulties with the plugs becoming coated with lead deposit. This was due to low charge temperatures under low power cruising conditions allowing concentrated pools of fuel to build up in the induction pipe without being vapourised, these then went into the cylinders in a gulp causing the plugs to lead up and so misfire. The drill to overcome this was simple: the engine rpm were increased which caused the charge temperature to rise thus vapourising and clearing out the collections of fuel either through the volute drain or directly into the cylinders. If this was done every ten minutes or so, the trouble was overcome. It was surprising how difficult it was to persuade the pilots of this!

By 1944 when we were really getting on top of the enemy and had complete air superiority, some air lines began operating again, although their passengers were mostly VIPs. Transport Command had greatly expanded and was carrying service personnel all over the world using the Avro York which was a development of the Lancaster, and also Liberators. It was important that the Yorks should be operated correctly from the engine handling point of view, in order to achieve maximum range. There were two schools on how best to achieve this and how to get the longest life out of the engines. BOAC were using the cruise control method which meant that power was reduced as the fuel was used up and the aeroplane became lighter, thus keeping the speed constant. We advocated using the constant power method which meant that as the fuel was used up and the aeroplane became lighter, it could climb higher and so the speed was increased. This method gave a shorter block to block time and because it kept up the charge temperature, it avoided the plug leading troubles.

Peter Birch and I made arrangements with Transport Command to fit a York up with a flowmeter and to fly out to Ceylon and back in order to demonstrate the effect of varying temperature on the fuel consumption and also to compare the two different types of operation. We obtained good results from this test and demonstrated that the constant power method was very satisfactory and so it was adopted in the Command. Trans Canada Airways, who were operating Lancastrians (another adaptation of the Lancaster) between Montreal and Prestwick, across the Atlantic also used this method.

Some of the Americans one came across did not seem to know too much about British engineering achievements. On one occasion, one of the American squadrons I spoke to didn't know the history of the Merlin Mustang and thought it was just another all-American fighter. Another time, when an RAF Lancaster flew over to the States and landed, its engine cowlings were removed for inspection and an American crew man remarked, 'I see you use Packard engines' and then went on to say 'So this is a Lancaster—over in Europe they call them Liberators!'

It was very pleasant meeting characters overseas whom one had known well in the early days of the war. People such as Duncan Smith, Ronnie Lees, Brian Kingcomb, Cocky Dundas, Paddy Barthropp and Mike Rook to mention just a few. In three years of war they had amassed the experience of half a life time. Most of them had now become Group Captains and had been highly decorated in the intervening period.

Stokey spent most of his time visiting the Mosquito squadrons at home. He was well qualified to do this as he had been operational on tactical bombers and had been involved with engine development with the Mosquito at Hucknall. The problems with the Navy were tackled by Dick Peach. He concentrated on visiting the carriers, giving advice on the operation of the Firefly, Barracuda, Seafires and Sea Hurricanes. He went up to the Arctic to witness some cold weather trials. These trials consisted not only of proving the equipment but also of the survival of the crew. He volunteered to be immersed in the sea in a survival suit. When they pulled him out he was nearly unconscious, so they gave him a half pint of Navy rum to revive him. This warmed him up so much that he wanted to repeat the test—or so the story goes! It showed that at least the Rolls-Royce reps were willing to try anything!

Rolls-Royce above all other companies appreciated the importance of ensuring that their products were being operated to the very best advantage, and we always received the full support of the management in our endeavours to accomplish this.

The Advent of the Jet 8

I first heard of a jet engine in 1941 when I attended a Research meeting at Belper chaired by Hives. Dr Griffiths had produced a drawing of a contra-flow gas turbine engine which at the time seemed to me to be very far fetched and impractical. At the same meeting Dr Hooker talked about the Whittle engine saying that they were having trouble with compressor development and that he had been asked to help—did Hives think he should? Hives typically remarked 'By all means help them. After all, you are the expert on superchargers and we are all in this together.'

It wasn't long after this that Hives saw which was the right line of development to follow and started negotiations to take over the Whittle engine to bring it to a production status; the Griffiths engine remained a research project. The Rover company was also interested in the Whittle engine, but it was decided that Rolls-Royce should take it on and that Rovers would take over the Rolls-Royce Merlin tank engine. This was obviously a wise and far-sighted move on Hives part and it was to ensure that Rolls-Royce would become world leaders in gas turbine technology—a position which they have maintained to this day.

A factory was opened up in Barnoldswick and Stanley Hooker was put in charge. Lombard came from Rovers and was made chief designer and a most enthusiastic and able team was formed, which produced remarkably efficient and rapid results. Once the jet engine became a practical proposition and developed to the stage of going into production, it became a subject of high priority. It was known that the Germans would shortly have squadrons operational with the Me 262, the Arado 234 and the Heinkel M-163, and so it became urgent to counter this. The development of the Whittle Welland to a practical, operational engine was the first task, soon followed by the Rolls-Royce Derwent 1 and 3. Once the Derwent was brought up to type-test standard, its early flight tests having proved satisfactory, the engine went into full scale production at Barnoldswick to match up with the Meteor production at Glosters. Development on the Nene engine, which was a scaled up version of the Derwent, designed to give 5,000 lb of thrust, was also begun at this time although it was not produced until the war was over. A whole host of engines were designed, some were built and others were just patented, but the whole spectrum of cycles was explored.

The initial Meteor flight tests were shared between the test pilots at Gloster, led by Eric Greenwood and John Grierson, and Group Captain Willie Wilson AFC and two bars, from RAE Farnborough. Concurrent with this, we at Rolls-Royce were allotted several Meteors for development flying to become accustomed to the different handling techniques.

Our first reaction was, how easy they were to fly, but how quickly the fuel was used up. This was normally no problem as there were so many aerodromes on which to land, but with the Meteor using kerosene, one had to get back either to Hucknall or Church Broughton to refuel. On take-off, the difference between the Meteor and the high-powered piston engine aircraft was very noticeable. The Spitfire for example were taken off with full rudder to prevent swing, and opposite aileron to keep the wing up and counteract the torque. This condition lasted only a few seconds before the aircraft fairly leapt into the air. The Derwent engines however were smooth, the aircraft accelerated slowly and there was no tendency to swing off the runway. When a speed of about 130 knots was reached the stick was eased back and nothing much seemed to happen—rather an insecure feeling due to the low thrust of the early engines. Another interesting factor when we began flying these aircraft was that with the much higher speed, it was easy to get lost! We used to enjoy 'beating up' any fighters we saw cruising around, especially the Mustangs from the American squadrons, for at this time there was a high level of security imposed on the jet development and nobody knew what they were: all the Mustang pilots knew was that here was a plane, much faster at low altitude than their own, and using no propellers!

The first squadron to be equipped with Gloster Meteors was 616 Squadron which was formed in 1944 at Manston from Willie Wilson's flight at Farnborough. The Commanding Officer was Jock McDowell DFM and bar. Mike Graves (who later joined us as a test pilot) was a Flight Commander and George Wilkes was one of the senior pilots; they both did valuable work in getting the squadron converted to jets. When the squadron was training, they were used against the flying bombs coming in up the Thames Estuary, a task which suited them because of their high speed at low altitude. I had, by now, done a considerable amount of flying in these aircraft and had learnt a lot about them. In order to increase their speed still more, I obtained a set of reduced diameter exhaust nozzles and had them fitted to one of the aircraft. I decided to test them before the squadron accepted them. On opening the throttles for take off, both engines surged violently as I accelerated along the runway. Due to this, I was not obtaining sufficient thrust to become airborne. As the end of the runway approached I eased back the throttles and immediately the banging stopped, and the thrust increased, enabling the Meteor to become airborne. We were learning the idiosyncrasies of these new engines all the time!

Whilst doing this, Willie Wilson was trying out the first attempt at 'afterburning' at Farnborough. The object was to gain extra thrust by burning fuel aft of the turbine. Hives, whilst visiting Farnborough went to have a look at it. After examining the rather crude fuel system, he made one of his classic,

The Derwent Meteor

and in this case rather prophetic, remarks: 'There must be more to it than just piddling petrol into the jet pipe and setting it alight!' A modern re-heat system as on the Spey Phantom today augments the thrust by 80 per cent and weighs and costs about one-fifth of the main engine! But these were early days and the jet engine appeared to be a simple power unit compared with the rather complex and highly developed piston engines of that time.

When 616 Squadron became fully operational, it was moved over to the German base, Quackenbrack. The squadron had been complaining of the frequent failure of the Michel thrust bearings and so Cyril Lovesey and I flew over to investigate. Cyril realised that only replacement by ball race would suffice and so we returned to put this in hand. An amusing incident took place after our return to Hucknall: we had brought back all sorts of 'goodies' for Christmas with no customs inspection or immigration clearance. Deteriorating weather conditions had forced us to leave Brussels in a hurry and we had been unable to notify the authorities that we were going direct to Hucknall. The day after our return Cyril was summoned to Hull by the authorities to explain why, he a civilian had re-entered the country in this fashion. I was in the clear as I was in RAF uniform. Cyril had a good story as usual, and started to explain that it was all the pilot's fault! The inspector asked the name of the pilot and when Cyril mentioned that it was me, the inspector replied 'Why didn't you say that before? I knew his father well when he was the Port Medical Officer at Tyneside, and we used to check all the foreigners together coming into Newcastle!'

The next engine to be built at Barnoldswick was the Derwent 5. This was most successful in the Meteor in which both Willie Wilson and Eric Greenwood broke world speed records, achieving speeds of 600 mph plus. This was followed by the Nene which went into the Vampire, the Hawker Seahawk, the Supermarine Swift and Attacker and a number of experimental prototypes. One was also fitted to a Lockheed Shooting Star. This nearly repeated the Mustang exercise: the installation was done in a very short space of time and over 1,000

The Nene Shooting Star

were built in Canada and the United States where Pratt and Whitney took a manufacturing licence—Rolls-Royce having built a factory in Montreal to build them. Grumman chose the Nene and later a scaled version, called the Tay (of 6,500 lb thrust), as the power unit for the Congar and Panther which had been ordered by the United States Navy.

The French also used the Nene and Tay; they were built under licence by Hispano-Suiza for all their experimental prototypes and for the production Dassault Oregon and Mystere. The Nene Vampire was also used in the French Air Force. The Russians came to Rolls-Royce to see the Derwent and Nene on invitation from Sir Stafford Cripps, the then President of the Board of Trade in the Socialist Government. He gave permission for them to purchase a number of engines against considerable opposition of opinion: it was feared they might copy them. Of course they did this and they built them in Russia in large numbers to be installed in the Mig 15 and 17. They didn't pay a licence fee either! The Americans were furious, rather naturally, as they were soon to be fighting them in Korea.

Hives now realised that the jet engine development and production had outgrown the factory at Barnoldswick. It was decided that the development on future experimental engines should remain there but that the production should be moved to the main works at Derby. It was essential that the main force of engineers became accustomed to the new design and manufacturing techniques of the turbine engine which was now rapidly superseding the piston engine. Stanley Hooker returned to Derby and was put in charge under Elliot. However, problems arose on compatability and so he left the company to become chief engineer at Bristols, who were just beginning to get involved with turbines. Adrian Lombard was put in charge and rapidly built up a first-class team of high-grade engineering talent with Cyril Lovesey as his deputy and in charge of all aero engine development. The following anecdote illustrates the comradeship of British engineers and the generous feeling of both Stanley Hooker and Hives, quite remarkable to some foreigners who cannot understand this 'cricketing' spirit when commercial competition was so keen.

It was no secret that Stanley Hooker and Hives had fallen out, hence the former leaving Rolls-Royce and joining the rival firm. One day when Hooker was sitting in his office at Bristol, late and alone pondering why the Proteus engine was in such serious trouble, his telephone rang and a voice said 'That you Stanley? Hs speaking. I hear you are in bad trouble with your engine.' Hooker wondered what was coming next! Hives went on to say, 'Just say the word and your old team will be on the next train to Bristol to assist if you need them.' This co-operation also happened in reverse and Bristol people went to Derby to help on some Rolls problems. The object behind this of course was to get the job right so that the customer, whether it was the RAF or the airlines, could be got out of trouble quickly, and to achieve this they were prepared to temporarily forget the competition between the two firms.

On the question of rivalry, the main competition at this time was between De Havillands and Rolls-Royce. Bristol came in rather late on the turbine scene. De Havilland under the leadership of Major Halford and Doctor Moult had designed the Goblin engine, the dominant feature of which was the single-sided impellor on its centrifugal blower which drew air direct from the forward facing air intake so that there was little interference. Rolls-Royce favoured the double sided impellor, which required a plenum chamber to collect the air which entered the engine on both sides of the impellor but which enabled the engine diameter to be kept smaller. Not much was known at this time concerning aerodynamics of air intakes, thus there was much scope for debate and salesmanship as to which was the better method. In practice the installed thrust of the Goblin was in the order of 3,500 lb which it also gave on the test bed, whereas the Nene gave 5,000 lb thrust on the test bed and only 3,500–4,000 lb when installed in the aircraft. The discrepancy was due to air intake losses. On the other hand the Nene was considerably smaller in diameter and so the nacelle drag was lower. When installed in the Vampire both engines gave similar top speeds, although the Nene climbed better. On faster aircraft the Nene gained an advantage as its smaller diameter and less drag was of greater importance.

This resultant competition between the two firms kept everybody engaged in sales promotion on their toes. It was highlighted when the RAF were preparing to equip a number of squadrons with the Vampire and the decision as

The Avon Canberra

to which engine, Nene or Goblin, would be selected had to be decided. We had two Vampires at Hucknall; one fitted with the Goblin and the other with the Nene, and we decided to invite the Director of Operational Requirements, Air Commodore Tom Pike and his Deputy Wing Commander, Dennis Crowley-Milling (whom you will recall was one of the last premium apprentices before he joined the RAF) to fly them both. This we believed would firmly instill into their minds the advantage of adopting the Nene engine.

Neither of these very able fighter pilots had much experience of flying jets. Dennis Crowley-Milling took off with the hood open—normal practice in current piston-engined fighters but not on jets—and we were most alarmed to see the hood fly off just as he became airborne! Fortunately this was without serious consequence, although we felt a little guilty about it happening in case we had forgotten to tell him during the briefing. The Goblin engined version was finally chosen, primarily due to the extra air intakes, known as 'elephant ears', they were required on the Nene but unfortunately caused buffeting in flight, thereby reducing the Mach limitation.

Aircraft and engine development were proceeding apace; smaller frontal area was a necessity as the speed of sound was approached to lead on to supersonic flight. It became necessary to abandon the centrifugal engine in favour of the axial flow type. This feature became the critical part of the engine as rapid advancement in technology was taking place. Armstrong-Siddley now came into the competition with the Sapphire engine.

Rolls-Royce had designed the Avon which featured an axial compressor; it was primarily intended for the new Canberra bomber. In its early stages the engine suffered severely from compressor stall, known as 'surging': the design of the air intake became critical to enable the engine to run satisfactorily under varying flight conditions. The Armstrong-Siddeley Sapphire was less adversely affected which gave it an advantage over the R-R Avon. There were a number of new aircraft projects on the drawing board so competition was intense as to which engine would be selected. Rolls-Royce redesigned the compressor on the lines of the Sapphire, thus covering the surge problem and this coupled to other technical advantages such as air-cooled turbine blades gave the Avon the edge. Aided by the great reputation and experience of Rolls-Royce, most of the aircraft manufacturers selected the Avon.

This sort of competition exists to this day; for example, by-pass engines versus straight jets, two- or even three-spool engines versus single shaft. In the design of engines for vertical lift aircraft, it was variable nozzle versus multiple super-light lift engines and so on. All this rivalry was very good for the industry as it drew the best efforts from everybody concerned whether they were engineers, sales staff or even ministerial civil servants and politicians. The customer, both aircraft manufacturers and operators always gained from this by being offered a better specification and performance through accelerated development.

With the Meteor beginning to reach the squadrons, there were understandably teething troubles to be overcome with the engines. This meant frequent visits by the top engineers to obtain first-hand information. It was during some of these visits, when I was the pilot, that we had some rather hairy incidents, which had fate been unkind might have set back engine development considerably! Perhaps Stanley Hooker or Adrian Lombard had a jinx or maybe they just wanted to press on regardless!

On one occasion, I flew a party of engineers from Hucknall down to Manston in our Oxford to attend a dinner to celebrate 616 Squadron becoming operational. We had a pleasant evening with the pilots and their wives and on the morrow prepared to leave for the North. The weather was really foul with low cloud and rain and bad visibility and I thought about delaying the departure. However, the weather was obviously not going to improve and as Lombard as usual was keen to press on, I let him persuade me to take off.

We became airborne and were in cloud at 100 feet. We turned north to cross the Thames Estuary and were flying between the sea and the cloud base, when about halfway across, the port engine lost power. I throttled it right back, but having non-feathering propellers we were only just able to maintain height on one engine. We managed to turn round on the reciprocal course and we tried to find Manston again. Everybody on board was now rather agitated, and all eyes were searching through the murk to try to find the coast line and the aerodrome. We skated over the hills with very little room to spare and found the runway lights. I flew in, in a position to do a 40° turn, lower the undercarriage and flaps and go straight in but just as I was in the middle of doing this a Mosquito loomed up right in front of me, having turned inside us from the left. It was in full fine pitch with a lot of throttle on, motoring in to land in front of us. We flew straight into his slipstream which tilted us up vertically on to our port wing tip with only about 100 feet in which to pick the wing up, there was no hope of going round again. By putting on full top rudder and full aileron and full power on the starboard engine we levelled out as we hit the runway but with a lot of drift on. Fortunately nothing broke and we were on the ground in one piece. It was a near thing!

We had to stay the night while they diagnosed the trouble, rectified the fault and made it serviceable for the next day. Our troubles were not over! We decided to go to the local in the village nearby called Plucksgutter. Tim Kendall, the Rolls-Royce representative at Manston, kindly offered to take us in his Rover car. We were going fast along the road behind another car when suddenly it swerved and something hit our radiator cap cutting it right off, and forcing it to land on our roof. Tim pulled up to see what had happened. It was quite simple but could have been lethal; both the car in front and ourselves had hit a steel cable which was strung across the road, it had broken the windscreen of the car in front but fortunately for us it had caught our radiator cap which had dissipated some of the energy and so our windscreen had remained intact. At this time most main roads adjacent to aerodromes had wires supported by pylons strung out along them to prevent German gliders from landing on them; this particular wire had become unshipped from its pylon and had been dangling across the road, hence the accident.

As if this wasn't enough for one night, Eric Greenwood and I were later awakened by the sirens warning us that flying bombs were coming up the Thames Estuary. We could see the flaming exhausts coming towards us and then they became invisible. This was the time to get under the bed as it meant that the engines had stopped and that there would therefore be a big bang any minute. Surely enough there was. Then it happened all over again, we saw the lights, they went out, but this time there was no bang. We couldn't understand it, so we got bored with it all and went back to sleep. In the morning we discovered that they hand't all been flying bombs but were mostly the local 'Swordfish' squadron doing night circuits and bumps. The lights were the navigation lights but the big bang was an unfortunate pilot who crashed on the end of the runway carrying a torpedo.

Next morning we took off and returned to Hucknall. The port engine lost power once more as we neared the aerodrome but all went well and we arrived safely just in time for lunch.

On another occasion when bringing Stanley Hooker and Witold Challier back to Hucknall from Barford St John near Banbury, where the early jet test flying was being carried out, we had quite a fright when taking off in the Airspeed Envoy. I wound the tail trim forward in the usual manner and opened the throttle and accelerated down the runway. I then eased the control column back and found that the nose would not come up! I wound the wheel back as quickly as I could but to no effect. Somehow I noticed that there were rather a lot of people in the cockpit and then realised the centre of gravity was too far forward. I yelled 'Get back into the cabin for God's sake we can't unstick!';

Stanley and Witold dashed back to the rear and then of course the nose came up suddenly as the tail trim was now too far back, however, we became airborne and trimmed the aircraft and off we went back to Hucknall. This episode became a standing joke with us all for many years and it was often told to visitors at dinner at Duffield Bank House. On such episodes hangs the fate of future events which reminds me of a story that Kurt Tank the designer of the famous German fighter FW-190 told me after the war.

He was flying a small Focke Wulf cabin monoplane from occupied France to Holland at the beginning of the war when he was attacked by a pair of Spitfires. His aircraft was damaged and he had to force land in a field, the Spitfires having departed. He remarked to me 'If those pilots had pressed home their attack they would have killed me and there would have been no FW-190!' How nice that would have been for the RAF! He then went on to suggest I should find out from the fighter command records who the pilots were—how arrogant can one be!

Hives decided to go for the South American market in a big way. The plan was for Stanley Hooker and Lindsay Dawson to go with him, they would give a series of lectures on turbine engines and jet propulsion to learned societies and Air Force establishments and so spread the word all around on the advantages and the future of this prime mover. They would also of course stress how Rolls-Royce had been responsible for much of the pioneering work and therefore knew more than most about the subject and generally point out the advantage of working with the company. Hives thought it would be a good idea if the three of them learned Spanish, so a teacher was found who gave them lessons for a few weeks prior to their departure. I believe Stanley and Lindsay can get by with the language, but the story goes that Hives was too busy to do his homework and got his secretary to do it for him!

The result of their expedition was highly successful: the Argentine ordered squadrons of Meteors with Derwent engines and also Lancaster Bombers with the Merlin. Kurt Tank, who was then working in the Argentine, also accepted a Derwent for his Pulqui.

Once the war was over, many of the fighter squadrons were disbanded and the volunteer and reserve pilots were demobilised. The Auxiliary Air Force however remained in being and carried on as before. The bomber squadrons remained active longer as they, together with Transport and Coastal Command, brought back troops and prisoners of war from the foreign war theatres. Aircraft and engine production was cut back from the maximum possible to a mere trickle. It was fortunate that the new prime mover, the turbine engine, had come in to supersede the piston engine at this time, as this meant that the RAF and the Royal Navy had to be re-equipped with new aircraft. Piston-engined planes were now becoming obsolescent and this meant that there would be funds available for developing and improving existing turbine engines, such as the Nene and Derwent. The gas turbine was in its infancy, and there was great scope for inventing new cycles; such as, bye-pass, turbo-propeller engines, axial compressors, etc., and also in all sorts of sizes. There was no worry therefore for the designers and engineers, but lots of headaches for the management and production planners. The machine tools for turbine engine manufacture were quite new, as also were the techniques to be learned.

Efforts were also being directed to putting Civil Aviation back on to its feet. There were no British airliners, so BOAC and BEA and many other airlines had to start from scratch. The Americans had had an eye to the future before the war ended, and because they had many more transport aircraft than Britain due to Britain's concentration on building combat aircraft, they were in an advantageous position to supply the Douglas DC-4 to any allied or foreign countries who wanted them. Britain only had the York to offer; this was a transport aircraft developed from the Lancaster bomber. Thus competition was stimulated between UK and USA to supply this urgent need. This was fortunate for the industry and it soon became an activity increasing in proportion to equal the size of the military market, both for domestic and export.

The first jet airliner: the Nene-Merlin Lancastrian

Things weren't too good for myself and my team however and soon Hives sent for me and said, 'When things get slack the work force must be cut back and the first people to go are the cause of the overheads, such people as pilots and liaison staff.' This sounded most gloomy, but he went on to say that it was necessary to go out into the world and get orders, and so he had decided to send my team out to appoint agents in Europe and try to pick up as much business as possible. The allied countries which had been liberated such as France, Denmark, Belgium and Holland had all been given Spitfires, Mosquitoes and Mustangs, etc., in order to build up new Air Forces, to train pilots and ground crews and to start overhaul shops and maintenance depots. We were told to go out and generally get to know how the various countries intended to develop concerning both military and civil aviation. He finished up by saying 'Don't come back empty handed. If you get kicked down the stairs, don't worry, just go back up the stairs so long as you get an order.' This was very different advice to that which had been given to me several years previously. Then he had told me to 'Go and find out what the customer wants; it is just as easy to give him what he wants, as what he doesn't want!' Circumstances were different now and I realised that in this case the customer was merely learning, Rolls-Royce needed the business and it was not a question of producing the best equipment for the front line squadrons. This doctrine was however quite new to us and I realised that we too had much to learn, for selling and commercial practice were quite outside our experience.

My team of test pilots was now down to two: Harry Bailey and Group Captain Rendell Stokes. Dick Peach had been killed flying a Canberra, Tony Martindale went back to developing motor-cars at Rolls-Royce, Crewe, and Peter Birch was chosen by Hives to go out to Australia to start an office out there. This he did with Alec Bailey, most successfully, building up an engine overhaul facility and an association with the Commonwealth Aircraft Company in Melbourne in which Rolls-Royce took a ten per cent share of the equity.

Stokey had been a fund of advice to us since he had joined the team. Most of it was good but some a little unorthodox, bless him. He was older than the rest of us and in many ways more experienced, and so we always listened to him even though we didn't always follow his council! We allotted him the Middle East territory: Turkey, Greece, Egypt and Portugal. Harry Bailey had Italy, Switzerland and France, and I took Scandinavia, Holland and Belgium. I decided to fly round my parish in the Proctor and took Tim Morice from Dunlops as a companion as he was also on a sales drive for his company. He and I had known each other for some time. He was a most remarkable character; having won the MC in the First World War and a DSO in the Second World War as an Air Field Commander in Sir Harry Broadhurst's 83 Group at the age of fifty-six! Apart from having a wonderful sense of humour he was a wise and lively companion.

I picked him up at Hendon and we flew on to Amsterdam in the Proctor where we spent our first night. It was all very pleasant and it was interesting seeing new countries in an emergent state so soon after the war and recent German occupation. Some of the girls in a night club showed us their concentration camp numbers tattooed on their arms, and told us rather gruesome stories of their experiences. There were many bicycles running around without tyres on; public services were gradually recovering. The one thing which was first class was the food; we in England were still on rationing although we had won the war, but in Amsterdam, at least in the expensive hotel in which we were staying, there was no shortage of meat, game and other delicacies, at least at a price.

We went on to Denmark, then Norway and Sweden. The Danes were very friendly, Norway we found rather austere as they had had a pretty rough time, but Sweden having been neutral was really elegant and gay. There was everything one could wish for in the shops; bright lights and oppulence could be seen everywhere. The women were well dressed and mostly very pretty; it was summer time and they were sun tanned, and taken altogether we were glad to be there!

We enjoyed some lavish entertainment partly because the word got around that we were looking for agents—and what could be better than being selected by Rolls-Royce or Dunlops? One chap came to see me in my bath! Another had a Prince on their board and so on. The agent we finally selected took us for a most enjoyable weekend cruise on a yacht in the archipelago; a memorable and happy occasion.

The only order I managed to collect was for one mile of phosphor-bronze strip for putting on the edges of skis. This was from the firm I intended to appoint as our agents, they had a small factory for making skis and had been unable to obtain this material during the war. When I returned to Derby and told Hives he said, 'Good God, we are not as badly off as all that.' However we did execute the order.

The Swedes had recently bought a number of Vampires fitted with the Goblin jet engine; these aeroplanes cost about £30,000 each, the agents commission had been the usual 5 per cent. The Swedish Government had taken exception to paying the resultant large sum of money to the De Havilland agent and had decreed that in future they would only agree to a much smaller rate of commission. Having selected the firm I wished to be appointed I negotiated with them to accept only 1¼ per cent when I told Hives how clever I had been in driving such a hard bargain he said, 'We can't have anyone working for us at such a small rate,' and promptly sent a note to Col Derby, the Aero Export Manager in the London Office, instructing him to sign a contract for 2½ per cent. Col Derby being a little out of touch with reality and thinking of pre-war rates finalised the contract at 5 per cent. This wasn't too bad when all we were selling was a few spare Merlin engines, but it wasn't long before the Swedes took an Avon licence and bought a large number of Avons for their Draken and Lansen fighters; engines costing around £30,000 each. This firm soon made a fortune. This was the sort of learning curve in commerce we were experiencing! It was all good experience and lots of fun too. We were enjoying our first taste of peace-time activity.

I did many trips round Scandinavia flying the Proctor and had a number of quite exciting experiences. My passenger on most of these occasions was my very good friend Bert Milward of airship R-100 fame! I had been looking for an agent in Holland but had not found anybody suitable. Bert had been the

Rolls-Royce representative there before the war when he was with the Dutch Air Force; he had married a talented Dutch girl Leicje, and was now the Sales Manager of Rotol Ltd a joint company owned by Rolls-Royce and Bristols who made propellers. When I suggested to Hives that he should be appointed as the agent he said 'You can't have Milward he's a black hat and umbrella man for Rotols'. Eventually after persisting over a period of time Hives agreed and Bert was appointed. It was the right decision as he boosted Rolls-Royce's products in Holland and promoted much business and generated a great deal of good will in that country.

On one occasion when we had flown out from England to Schipol intending to go on to Copenhagen after refuelling, we had borrowed a strip map from a KLM pilot which gave a narrow corridor and a red pencil line to follow. The weather was bad and after take off we found the visibility was down to half a mile and the cloud base 2–3,000 feet. After about an hour's flight we had drifted off the strip map and were lost. The radio was of little use as the channels we were using were no good on the continent and so we just flew on hopefully thinking we would strike the river Elbe. We found a river but did not recognise it, Bert produced a small pocket diary with a map of Europe in it and this showed that the river was the Weser and that the next one up would be the Elbe. By the time we found it we were running short of petrol and the visibility was very poor indeed. Bert was not in the least nervous or didn't seem to be! I had landed at Hamburg once before and so remembered that the Aerodrome Fuhelsbutel was north of the town; I really felt that we wouldn't have enough fuel to make it and I was getting very apprehensive. We finally landed safely, but only after flying up the river to Hamburg and then crossing the city at nought feet with the fuel gauge showing zero!

Stokey was equally making his name as a flying diplomat in Turkey. He had had a narrow escape when delivering a Mosquito to Ankara. Having arrived safely, he had persuaded the Turkish Commander in Chief to fly with him and experience a demonstration. All went well for a time until Stokey demonstrated single-engined flying, by feathering one of the propellers. When he attempted to unfeather he found the battery was flat and it wouldn't work! He then said he would demonstrate a single-engine landing, which he did. A Mosquito tends to swing on landing and when Stokey tried to correct the swing by applying the brakes, alas there was no air pressure! The result was that the swing became uncontrolled, the aeroplane swung round and the undercarriage collapsed resulting in serious damage. Luckily neither of them was hurt. Stokey, worried about who would foot the bill, sent a cable back to Derby saying 'slight landing accident aircraft damaged ignore rumours'. This was received with considerable laughter and the cable was duly pasted in the 'linebook'.

Our liaison team finally split up now; Stokey joined the sales department under Col Fell at Derby, Harry Bailey went back to full time test flying at Hucknall, later to become the Chief Test Pilot and I was appointed to the London office to replace Col Derby who had recently retired as Aero-Export Manager and RAF liaison contact. Bill Lappin, who had been Hives personal assistant and high level contact with the Services and Ministries, was growing older and so I was to prepare to take over from him when the time came.

When I left Hucknall the beautiful walnut billiard table made by Burroughs and Watts which Father had given to my brother and myself and which had been so admired in my Nottingham house, was now surplus to our needs. I decided to donate it to Rolls-Royce at Hucknall to be used in the welfare hall. I wanted it to be a token of appreciation to all my work mates who had been associated with me during this rather important phase of my career. From now on I would just be an infrequent visitor to Hucknall, going there when the job in hand demanded. It was with considerable regret that I said farewell, but life moves on, circumstances demand changes and after all it was promotion; so off I went to the big City and a challenge with much to be learnt.

Hives Becomes a Baron 9

We were all delighted when the news came through on 16 June 1950 that Hives had been honoured by His Majesty King George VI and been raised to the Peerage. Previously when he had been made a Companion of Honour, a unique distinction for an engineer, we had all felt very proud of him; but now we really felt he had at last been given due recognition for his outstanding contribution to the war effort and for advancing the prestige of Great Britain in the international field of advanced engineering and technology.

A celebration was held at Derby to commemorate this event; it was held in the staff canteen and took the form of a luncheon. I will quote from two speeches after the lunch as these sum up most lucidly the driving force and spirit which motivated the performance of the company and which had been constantly stimulated by the personality of Hives. After receiving the honour he issued a message to the workers as follows: 'It is my wish that the great honour which has been conferred upon me by His Majesty should be shared by all Rolls-Royce workers.'

Mr Swift, the general works manager and an old colleague, spoke thus:

> 'We are here tonight to celebrate the elevation of Mr Hives to the Peerage, an honour well merited and long overdue, and it is a great occasion for the Company, the town, the industry and his friends and colleagues, and sets the seal on a distinguished career. During the forty years I have known Hs, he has always been in the forefront of Rolls-Royce achievements, even in the early days of this factory Hs as a tester was always striving for quieter cars, and Sft as a charge-hand fitter on axles was always doing his best to please him.
>
> 'Later in life another of Hs achievements was the London/ Edinburgh trial which ended with a speed test on Brooklands of 100 mph on a standard car already world famous. In those days Hs was chief of a very small Experimental Department, directly under the control of the late Sir Henry Royce.
>
> 'In the 1914/18 war Hs was responsible for the development of the Rolls-Royce Aero engines, Eagle, Falcon and Hawk. For these and other outstanding services during the war he was awarded the

MBE. Later came the Atlantic crossing. He had a finger in this pie too. Then came the well known success on the Schneider Trophy races. Hs was also busy with motor car development. The Silver Ghost went. The Phantom and Wraith being born. Subsequently the Rolls-Bentley when the Bentley company was taken over by Rolls-Royce.

'Now during the idle periods the Kestrel aero engine was designed and developed and this engine put into production on then modern lines. The next serious and most important stage was when the Nazi war clouds began to gather. Rolls-Royce was asked to tackle an expansion programme of large, modern and efficient aero engines, and it was Hs who shouldered the responsibility of developing the Merlin, which as is only too well known engined the Spitfires and Hurricanes etc., in the Battle of Britain which put us on the right road to victory, otherwise this honour celebration might never have taken place at all.

'Just about this time, advancement again followed the heel of achievement, and Hs in 1936 succeeded the late Arthur Wormald as General Works Manager, and it was under Hs personal example and leadership that Rolls-Royce, by common assent became, as far as aero engines were concerned, the principle factor in the winning of the Second Great War. During this period, Hs was offered a Knighthood, but refused such an Honour whilst Britain was striving to live and might go under.

'After final victory was assured Hs was made a Companion of Honour, a magnificent acknowledgement of his personal achievements and leadership in the field of aeronautical engineering.

'When he joined the company in 1908, it was an era in which men accepted inequality of opportunity as a challenge—a challenge to be met by enterprise and resource and Hs met it in the fashion we know so well, hard work and honest endeavour. Never have I met a man who liked work so much, we all remember his battle cry "work till it hurts", another one "Work is the best fun on earth providing there is no bitterness and no financial grabbing".

'There have been other outstanding personalities over the years I have been associated with Rolls-Royce, but surely none greater than Hs. Great because he had the strength of character and an indomitable purpose to work for an ideal, well exemplified in the *Magic of a Name* and it is in keeping with such a man that so many of his old colleagues and men long in the service of the company are here tonight, they have all helped, however humble their task. I don't think anyone, workers or Government ever doubted that, we, as a team, would do our share in the fight and that we would do more than keep pace with the enemy. This team work was an outstanding tribute to his fine quality of leadership and to his tremendous breadth of vision. The RAF got the engines in the quantities and quality they wanted. The success of our first efforts was assured by the unselfish devotion to their duty of the Company's servants, none more so that Hs who despite his claim to be "only a mechanic" did nevertheless set an example and a tempo in the true sense of the word. His was the responsibility not only to provide the quantities, but to see that quality remained the watchword, and this honour is a vote of confidence from our fighting units, the RAF and others. He had the courage and ability to stick to an ideal under the greatest stress and to navigate this organisation, despite its temporary vast and unwieldy expansion, strictly in accordance with traditions proven in peace time and indispensable in war.

'Never has an award of His Majesty been so richly deserved and seldom have they been received with greater modesty and humility.

'To those who will have to continue this policy and tradition, I say he has laid a fine foundation for the future and this responsibility must be accepted when the time comes for him to relinquish the reins.'

Mr E. W. Hives then replied:

'I cannot begin to tell you how much I appreciate you all giving up your time to come here tonight.

'I know and you all know, that the Honour which has been conferred on me is a national recognition to the Rolls-Royce Company, and the Company includes all the workers, past and present, and especially includes people who are here tonight, because I can see around me the people who have given of their best to build up our reputation.

'I have been deeply touched by the little notes of congratulation and praise which I have received from all sections. The messages I have received confirm that we are a very happy organisation.

'You have often heard me say that I consider work to be the best fun on earth, and working for Rolls-Royce is even better.

'I think considering our numbers (over 24,000) and considering how we are separated by long distances, that there is a wonderful spirit of pride and loyalty, and we do not suffer from cliques or factions. To me, that is something we must always cherish and maintain.

'I can remember the time in the Derby Works when everybody walked about in fear and trembling of the Secretary! The Accountants were looked upon as something unclean, and if you wanted to know the cost of a part made in the Derby Works the only fellow who could tell you was somebody sitting in an office in Conduit Street.

Lord Hives

We have improved since then! You will find the Accountants are quite human. The proof of that is that they can even make mistakes, and better still—you can get them to admit it.

'There was a time when the Directors sat in solemn conclave and held their meetings in London. On the rare occasions when they visited Derby, meetings were held in the Midland Hotel. It has now been agreed that London is a relatively unimportant place compared with either Derby, Crewe or Glasgow.'

'We are fortunate in our shareholders. We must never overlook the fact that the company belongs to the shareholders. If enough of them got together they could undoubtedly give me the sack! but as the shares are so widely held there is no cause for alarm.

'When they talk of the freezing of wages, Rolls-Royce dividends to the shareholders have been frozen for 11 years.

'When we rejoice at the success of the company at the present time we must never overlook that it has only been made possible by the magnificent work of the people who laid down the ideals and built up the tradition of Rolls-Royce. In these we include Sir Henry Royce, Claude Johnson, Basil Johnson, Arthur Wormald and Sir Arthur Sidgreaves.'

'It has been my policy faithfully to follow the ideals which were maintained by the founders of this great firm. It is difficult to go wrong on the engineering and technical side if we follow the policy laid down by Sir Henry.'

'I would like to acknowledge the kindly co-operation we have always received from the workers in the Unions. I often boast to my friends that Rolls-Royce was one of the first companies in the country to recognise Shop Stewards and to make use of a full time Convenor. I certainly think we hold one record and that was an incident during the war when the Shop Stewards and the Management on one side argued a case on Women's Wages against both the Employers' Federation and the Union Executives plus the Ministry of Labour. Needless to say we won our point.

'One of the great changes in our business since the end of the War is the extension into foreign markets. There are few countries which are not our customers now. We have not only created new markets for our products but we have created new friends. In all parts of the world where our engines are either being manufactured under licence or being used, we have established a wonderful relationship. Personally, I place tremendous value on making your customers into your friends.

'One of the points which impressed me on my trips abroad was the appreciation there was for the moral business standard of the Rolls-Royce company: we must never allow a question of profit to jeopardise this position.

'Many people have remarked that the Honour is greatly overdue. I do not think I am giving any secrets away now when I say that I was pressed to accept a title in the early days of the war, and later, but I was a "conscientious objector" to accepting any Honour which was available to the men in the Forces who were making great personal sacrifices. It was because of my objections that I was awarded the Companion of Honour which came within my specification, because it was not awarded to the Services. It is generally given to the Church, to Poets and Men of Letters, with of course, Churchill and Atlee. I used to boast that I was the only plumber in the Union but this was disputed by Mr Essington Lewis, the Chairman of the Broken Hill Company in Australia who was awarded it the same year as myself and was a brother plumber!

'I have had numerous enquiries as regards what title I am going to adopt. The decision is a simple one, it will be the Rt Hon, Lord Hives of Duffield, in the County of Derby, possessing no estates or lands, I see no justification for making things difficult by changing my name.

'It is my intention to maintain this Honour with respect and dignity, because I look upon it as a trust earned by all the grand people who have built up this company.'

Certain cogent references in these speeches are worthy of serious thought today: how many men would refuse a knighthood when offered, rather than wait until the job is done? But Lord Hives in the true teaching of the Bible did so, being a humble man, and he reaped his reward in the form of the Honour which he accepted not for himself but on behalf of his workmates. This truly is the spirit of greatness.

There are passages in Lord Hives' speech which merit deep consideration today, for example 'I can remember the time in the Derby works when everybody walked about in fear of the Secretary!' 'The Accountants were looked upon as something unclean, and if you wanted to know the cost of a part made in the works, the only fellow who could tell you was somebody sitting in an office in London.' Also 'There was a time when the Directors sat in solemn conclave and held all their meetings in London.' Perhaps one of the problems of Rolls-Royce (1971) Ltd is that it has turned full circle?

Shortly after this, Lord Hives began to delegate the daily running of the company to his chosen successors; J. D. Pearson became Managing Director (later to become Chief Executive and Chairman of the Company) and Adrian Lombard replaced A. G. Elliot on retirement as Chief Engineer.

It is interesting to ponder on the fact that the Rolls-Royce Company lasted for 67 years without losing its identity as a self-contained public company, and even now the motor car division goes on, and maybe the nationalised Aero Engine factories could return to public ownership. If one considers all the other companies in Aerospace, they have all either merged or been taken over and thus lost their basic identity and peculiar attractions.

Having understood the ideals and the philosophy which gave Rolls-Royce its unique place in the regard and affection of its customers as described by Hives in his speech; one wonders if they can be retained by the new management in control of the nationalised company. Let us hope the spirit of Rolls-Royce will continue after such a drastic upheaval as becoming bankrupt.

London Office 10

I went to London to take up my new appointment in 1947. Before I left Derby Hives gave me some sage words of advice on how to represent the company; he informed me that I would become a member of the SBAC export committee, that I should entertain when necessary and that I should always be aware that I was representing the company image. I decided not to move my home from Nottingham, as I thought that if I lived at the RAF club and the Royal Aero club during the week and went home at weekends I would be able to get closer to the hub of things at the Ministries and be able to spend the evenings meeting people.

This scheme worked well; I was able to spend each Monday at Derby in the works to keep in close touch with current engine policy and the people in the departments with whom I had to have close association. A mistake in the past had been that the London office had been far too remote from the factory and that this had made people suspect the accuracy of their information. It was made clear to me that this must not be perpetuated. After all, the brains, the drive and the authority was centred at the works and it was important that the 'thinking' and policy should be passed to the ministries undiluted.

I was most fortunate in inheriting Col Derby's secretary Marcia Walker; she remained with me for twenty years. She was always most loyal and helpful in every way and I don't think we ever had a cross word. We remain good friends to this day. She left to be married and it was only then I realised how spoilt I had been and that I had never really appreciated how valuable a good secretary can be!

I found the London office strange at first after spending most of my time flying. I tried to keep up my flying by going to Hucknall once a month to fly a Meteor, but I found that it was a waste of time. I became neither one thing nor another: my flying proficiency fell off, and the time spent at Hucknall was time lost learning a new job in London. I finally decided to abandon active piloting after an incident in a Meteor which finally brought it home to me; the circumstance being a double engine failure and a thunder storm sitting over the aerodrome when I was forced landing. I managed to get one of the engines going and landed in bad visibility without mishap, but realised I was getting rusty and out of practice and so the lesson was learned.

I found my time in London fully occupied by keeping in touch with new operational requirements and then visiting all the aircraft firms who were competing, to offer them the latest brochures on Rolls-Royce engines. At this time, before amalgamations took place, there were some nineteen aeroplane designing and manufacturing firms working on government contracts. It will be of interest, I think to name them as many have now lost their identity in the mergers that have since taken place. They were: Hawkers, Glosters, Vickers, Supermarines, Armstrong-Whitworth, Shorts, Saunders-Roe, Fairey, Westlands, Blackburns, Avro, English Electric, Handley-Page, De Havilland, Phillips and Powis (Miles), Percival, Boulton and Paul, Bristol and Scottish Aviation. These famous firms were all highly individual, this largely stemming from the personality of their founders. They also reflected the characteristics of the part of the country in which they were situated. Avro and Blackburn were from Lancashire and Yorkshire; Bristol, Westland and Gloster were based in the west country and the rather charming flying boat people, Supermarine, and Saunders-Roe came from Hampshire and the Isle of Wight. The London and home counties firms were perhaps more aggressive and exhibited more of a sense of urgency or perhaps the need to survive than the others.

My new job was to involve a great deal of travelling. When I had been doing my job of flying liaison, I had either had a personal Hurricane, a Spitfire, a Mustang or a Proctor. Now that I was to give up flying, I felt that I should follow in Bill Lappin's footsteps and have a Bentley.

This was not going to be easy as there was protocol to be observed; Lappin was the only non member of the board to be so privileged and he was very close to the 'old man' and so in a unique position. However I went along one day to see Hives and I asked him if the company would share the cost of a second-hand Bentley with me. His reply was typical, 'You don't want to buy a second-hand car from the dealers, just hang on a moment.' He then rang up Doc Smith, who was the Managing Director at the Glasgow factory, and said, 'Send 8-B-5 down to Conduit Street tomorrow as I want it there for another job', and that is how I came by my first Bentley! 8-B-5 was an experimental pre-war car which had been used during the war by Hives himself and by various high ranking Air Force Officers.

It was valuable experience getting to know the chief personalities in all the firms and to become accepted by them. It was quite character building too, particularly when in confrontation with Sir Sidney Camm, Joe Smith and Sir Roy Dobson. They all had great respect for Hives, Lovesey and Elliot, and as their representative one was well received. George Edwards, the Chief Engineer at Vickers was becoming prominent at this time with the Viscount and later the Valiant which was the first of the V bombers to go into production. There was a very close collaboration between he and Lord Hives, Cyril Lovesey and Bill Lappin on these two projects, and a close relationship was formed. George later received a knighthood and many other honours during his exceptional career, where he has attained the distinction of being the best-known personality in the British Aero Space Industry. The aircraft firms all accepted that Rolls-Royce were experts on engine design and development and that they owed much of the improvement in performance to these factors. Most of them were critical when aerodynamic advice was proffered on such things as air intakes where there was an area of interfusion. This became difficult to resolve at first; Hives in his inimitable style used to say 'We are engine builders and know nothing about aeroplanes.' This at one stroke used to curb the arrogance of the younger and more progressive engineers and also used to please the aircraft designers. This over simplification disappeared as the flight test section at Hucknall developed and the interface between engine installations, particularly on gas turbines, became more involved and mutually reliant one upon another.

The European industry was becoming re-established once more after the dislocation caused by the occupation during the war and so one had to visit them also to keep them informed on Rolls-Royce engine developments. There was Fokker in Holland, Dassault (formally Marcel Bloch), Sud Oust, Sud Est, Nord and Breguet in France. Again this broadened ones outlook as these firms all had their individual Prima Donnas one had to cope with.

Now that the Meteor and the Vampire were entering service in the RAF, the question of a suitable trainer aircraft became important as new problems arose due to the different characteristics of jet aircraft. It was thought by the Ministry planners that a turbo-propeller aircraft would be the best compromise and would suffice for training both jet and piston engined aircraft. Two projects were proposed using the Rolls-Royce Dart engine; one by Boulton and Paul and the other by Avro.

Tony Martindale, who had had a lot of experience of both instructing and experimental jet flying and who worked with me at Hucknall as a member of my liaison team, supported me in disagreeing with the official policy. We wrote a paper saying that the next trainer should be a jet and not a propjet or a piston-engined aircraft. We saw the danger of De Havillands converting the Vampire into a trainer and so finding a large market for the Goblin engine at the expense of both the Rolls-Royce Derwent and Nene.

We approached Miles at Reading as we thought they were the best firm to design a trainer round the Derwent engine; they had had a deal of success in the trainer field during the war and had built hundreds of Masters and Magisters. The Miles brothers, Fred and George, and Heal, the chief draughtsman soon produced an attractive design.

The problem then arose how to launch it? There was no operational requirement as the policy was not in favour of a pure jet and so no money for its development would be forthcoming from the Government. I told Hives what we had in mind and persuaded him to visit Miles to look at the mock up and discuss the possibility of going along with them on a private venture and fund the project jointly. I flew down to Reading in our Miles Monarch and met Hives, who was driven down in his experimental eight-cylinder Rolls-Royce. We had a meeting and agreed it was an interesting project. It was suggested that the cost might be shared, but it seemed that Miles did not have the cash, and wanted Rolls-Royce to take some of their equity. This Hives would not do as he had had experience of this before the war, and the rest of the industry had taken exception to this as they preferred an engine company not to be committed to any one aircraft firm. I motored back to Derby with Hives and tried hard to persuade him to further the project, but he thought it too risky, particularly as he would get the Dart engine contract for either the Avro Athena or the Boulton and Paul.

Our next step was to discuss the situation with Dorey, my boss at Hucknall, and see if he would allow us to get Tom Kerry, the Rolls-Royce installation designer, to do a project and design a jet trainer so that I could interest Fokkers in Holland to produce one. The reasoning behind this was that Fokkers would want to start aeroplane design again now that the war was over, and they might as well start with a jet. Hives approved of this suggestion and so we went over to Holland to try and persuade them to do this. Assisted by Bert Milward, who had recently become the Rolls-Royce representative, we discussed the project with Piet Vos, Mr Baling and Van Meerten. Piet Vos was a good friend of ours as we had known him during the war in the UK. He had been at Fokkers before the war, and when the Germans had invaded Holland, he had appropriated a plane and escaped to England.

These gentlemen studied our drawings and listened to our philosophy that a jet trainer was the right type to train jet pilots! Alas Government approval would have to be obtained before money could be spent. This meant we would have to persuade Professor Van de Maas of the Government Aircraft Laboratories to agree. He too saw the advantage of the Dutch cutting their teeth on post-war modern jet aircraft by producing a trainer. The Dutch Air Force too had to be convinced, so this meant talking to General Aler; he too agreed and so Fokkers started to do a design. This became known as the S-14; Rolls-Royce agreed to supply a Derwent engine for the first prototype and so the project was duly launched. The prototype performed well but was considered to be under powered with the Derwent and so the larger Nene engine was fitted. This went into production and became the standard trainer for the Dutch Air Force. We tried to persuade the RAF to do likewise, but they would not grasp the jet trainer philosophy at this time. They did have a change of thought and

The Fokker S-14; the Dutch jet trainer with the Rolls-Royce Nene engine.

abandoned the turbo-prop and reverted to the piston engine; so the good old Merlin was put into the Boulton and Paul aircraft, and became the Balliol. It is interesting to note at this stage that Hives was very fair with me, and while pressing on with the Merlin trainer as official company policy, he did not want to discourage me by stopping me from trying to get the jet going, although the two were in conflict! His note to me, scrawled while he was at London Airport about to go to America, was typical of him and showed just what made him so popular with his assistants. It said:

> 'We have decided to back the Merlin Trainer. I want to make sure we do not find the two sections competing with each other. We have also no interest in aircraft design. This is not to discourage you but hold your horses until the Merlin Trainer is decided.'

While all this had been going on De Havillands had not been idle and had come to the conclusion that the Vampire could be developed as a trainer and so they followed this up by producing one and going to work on the Ministry to issue new requirements. The Vampire was then adopted and sold in large quantities both for the RAF and Royal Navy and many foreign Air Forces—just what we feared might happen.

The Fokker S-14 and the Vampire were now in competition for export orders; the S-14 very nearly had a bonanza in America. I was told by Dick Horner, who was Assistant Secretary to the USAF, that had it been available a year earlier it would have been chosen as the USAF trainer and built under licence by Fairchild. This would have meant an awful lot of Rolls-Royce engines also under licence being built by Pratt and Whitney. The irony of this was that the S-14 would have been available earlier had not Fokkers delayed its production by giving priority to a twin-engined piston trainer, which I believe never entered service.

I look back on these days of the advent of the Jet Trainer with pleasure; it was an interesting campaign although only partially successful and it gave one considerable experience at a time when all aircraft industries were in the stages of reconstruction after the war. One had the opportunity of meeting the up and coming engineers of many nationalities, and making many friends and travelling to interesting places. All the time one felt that one was being supported by the company and in particular the 'Boss' who was watching to see how you were going to turn out. The freedom of action and being allowed to get on with it was a great encouragement. Hives used to give good advice now and then to keep one on the right lines; not allowing one to stray too far. I remember that when I was getting rather one-track minded on the Jet Trainer to the exclusion of other projects, he told me to ease up a bit as 'one must not get too clearly associated with any one aeroplane or engine as one got type cast'. We had to preserve the broad outlook and see each project as part of a bigger jigsaw. One day Lord Hives walked into my office in Conduit Street, sat down and began to chat about things in general. He said 'You know "Lom" (Lombard) has a remarkable flair for engineering. He is the best engineer in

Mr Goodmanson, Lord Hives and the author at the Svensk Flygmotor factory, Sweden.

Britain, and of course that means the World.' A year elapsed before I told Lombard about this; we were in the States together, visiting Wright Field; I suddenly remembered the remark and mentioned it. He made me repeat it several times, because Hives was so sparing with his praise that when he did pay a compliment he really meant it, and it in turn meant a lot to us.

It was around this time I received a pleasant surprise in the form of a memo from the Financial Director. It said: 'We have come to the conclusion that it is time that you had a new motor car. I am writing to LS asking him to allocate a new car to you. Will you please speak to him or JS with a view to bringing this through as quickly as possible, because we want your present car for another purpose.' This was a terrific thrill. Incidentally, the other purpose for which the car was required was to be broken up!

Lord Hives decided to visit Scandinavia and as this was my parish he took me with him. I prepared the visit and made the appointments with the agents and the Commanders-in-Chief of the Air Forces and the senior industrialists. We were trying to penetrate the Swedish market and sell them the Avon engine licence. De Havillands had already been successful in selling the Goblin and Ghost licence, but now a new generation of fighters was being designed which required an axial-flow engine. The visit turned out to be both successul and enjoyable. On the way there we stopped in Denmark and in conjunction with T D M Robertson (General Manager) of Hawkers and Bill Humble (Chief Test Pilot) tried to persuade the Danes to buy Hunters which they eventually did. Mogens Harting our agent (selected on my original visit) did good work in bringing this about. He was also agent for Hastings, he looked after us very well and became a good friend. Lord Hives was an easy and pleasant companion with whom to travel and as usual very popular with everybody. I was exhausted at the end of the tour and wondered if I had passed muster. One rather had the feeling one was on trial and perhaps the future was at stake.

At this time (1947) we were in keen competition with De Havillands who were trying to capture the European market for the first jet fighter, with the Vampire to replace their piston-engined aircraft which consisted of RAF surplus Spitfires. We were supporting the Meteor which had two Derwent engines. Both types were in service with the RAF, so from that point of view neither had the advantage.

Both De Havilland and Gloster owned a demonstration model of their aircraft. Both companies organised sales tours round Europe. I was a member of the Gloster/Rolls-Royce team. Digger Coates-Preedy, one of the Gloster test pilots, was chosen to fly and demonstrate the Meteor, while Eric Greenwood, recently retired from being Chief Test Pilot, and myself flew the De Havilland Rapide which carried the ground staff, the luggage, spares and sales literature. We toured first Holland, Denmark, Norway and Sweden and finally Belgium.

The demonstration Gloster Meteor Mk IV with the Rolls-Royce Derwent engines.

The De Havilland team were also following a similar route. In spite of the sales rivalry we were all good friends, Dick Blyth the De Havilland pilot was always full of fun and good humour. The trip was enjoyable and it was pleasant to meet so many charming people of the various nationalities: King Haakon in Norway, Prince Karl in Sweden and Ministers, business men and Senior Officers in each country. It was in Belgium that we had our most exciting and frustrating demonstration, when hopes were dashed, then revived whilst our fortune hung in the balance. A joint demonstration of the two aircraft had been arranged by the Belgian authorities at Evere Airport, Brussels. Our Meteor was demonstrated first, flown well by Coates-Preedy. Dickie Blyth soon followed, putting up a neat and tidy performance, but having a less powerful aircraft, he flew in a less impressive way. A Belgian pilot was then briefed to fly the Meteor, he took off but was slow to retract the undercarriage, the speed built up and the nose wheel failed to fold back. He then went on to do violent manoeuvres with the undercarriage falling in and out according to the attitude of the aircraft. He finally landed with the wheels up, swinging off the runway and looking very derelict as the fuselage had buckled under the unusual stresses. Our hopes of a sale dropped to zero. Then the Vampire took off with a well-known Belgian pilot flying it; he did exactly the same thing; our hopes rose! and he too belly landed finishing up alongside the striken Meteor. We were now quits. The final sales result of our tour was a surprise; all the countries purchased Meteors except Norway and Sweden who chose the Vampire.

The Royal Air Force was now fully equipped with Meteors and Vampires so also was the Royal Auxiliary Air Force which was well trained and fully able to cope with the operational efficiency demanded. Mike Birkin was the Honorary Air Commodore and Inspector of the RAAF, having been CO of 504 squadron. He was very popular and under his guidance the RAAF was efficient. Alas, time was running out for the volunteers; it was not considered practical nor economic by the Air Staff Council for the RAAF to be equipped with either the Hunter or the Javelin, when they superseded the Meteors and Vampires and so the Force was disbanded. A very sad day after such a fine record of service.

The reader will recall my frequent visits and close liaison with the Air Fighting Development Unit at Duxford where the Merlin Mustang project was born. This establishment grew larger and incorporated the Fighter Leaders School, Tactical Trials Unit etc, and in 1945–6 became known as the Central Fighter Establishment (CFE). After being based at Tangmere it finally settled at West Raynham in Norfolk and became the show piece of Fighter Command. It attracted all the most experienced and well-known fighter pilots and was in turn commanded by such famous officers as Dick Aitcherley, George Harvey,

The test pilots' convention at CFE, named from the left. Back row:
1. Boxer. 2. Wright. 3. Binks. 4. Unknown. 5. Bird-Wilson. 6. E. Greenwood. 7. J. Cunningham. 8. R. Beaumont. 9. B. Bedford. 10. H. Satterley. 11. Vaughan-Fowler. 12. J. Derry. 13. P. Crisham. 14. Unknown. 15. B. Waterton. 16. D. Morgan. 17. Author. 18. Unknown. 19. B. Tuck. 20. J. Heyworth. 21. G. Tuttle. 22. Unknown. 23. P. Wykham. 24. N. Duke. 25. Unknown. 26. J. Lapsley. 27. B. Drake. 28. J. Kent. 29, 30. Unknown. 31. B. Oxspring. 32. F. Rosier. 33. Unknown.

Paddy Crisham, John Grandy, Fred Rosier, Geoffrey Millington and finally Bill Tacon. The staff pilots included all the fighter 'aces' Birdie-Wilson, Bob Tuck, Hawk Wells, Roly Beaumont, Bobby Oxspring, Dennis Crowley Milling, John Lapsley, Mike Giddings and Pete Brothers to name just a few.

Most of these pilots one had known as pilot officers at the outbreak of the war or even before and now they were all famous and becoming highly promoted before going on in their careers, many of them finally achieving air rank. They were a stimulating bunch of chaps and vastly experienced in aerial warfare and tactics. It seemed to me that their knowledge was of paramount importance and necessary to help in the successful development of the next generation of fighter aircraft. I discussed with John Cunningham and Eric Greenwood the idea of starting a Test Pilots convention to be held at CFE to exchange ideas on fighter development; they too thought it a worthwhile scheme and so I approached Paddy Grisham the Commandant to see if he would agree. Paddy of course was most enthusiastic and invited me to bring up a team of test pilots from all the fighter firms for a discussion on their latest fighter projects, nearly all of which were using Rolls-Royce engines. Rolls-Royce would thus get the benefit not only of the RAF's views but of the aircraft firms also.

Our team consisted of John Cunningham and John Derry from De Havilland, Mike Lithgow and Dave Morgan from Supermarines, Neville Duke from Hawkers, Roly Beaumont from English Electric, Eric Greenwood and Bill Waterton from Glosters and myself from Rolls-Royce. Discussions took place on all aspects of operations and tactics with the CFE staff some of whom had been on exchange in America and so knew much about US Air Force developments. Much valuable information was gained and taken back to the designers and engineers and used in the development of the new aeroplanes coming alone. After the meeting we entertained the RAF pilots to dinner, and further shop talk went on until late in the night.

After this first and highly successful convention we repeated them twice yearly. The chief designers wished to be included and so it developed into a large affair, this attracted the Operational Requirements Branch from the Ministry and so it grew to become a high level annual meeting with a guest night given by the RAF. Bob Tuck was one of the most colourful characters at this time; I first met him before the war when he was in a Spitfire squadron at Hornchurch and then later when he was leading a Hurricane squadron at Coltishall. He was one of the highest scoring fighter pilots before being shot down and taken prisoner, eventually escaping.

I mention all this as an example of the importance of getting alongside the customer and promoting good will and a close understanding of each other's problems; the urge to do this was instilled into one as part of the Rolls-Royce heritage and it paid off handsomely. It was a sad day when finally the establishment was disbanded in order to save money on the Air Force budget. It had had such a humble beginning as a detachment from 111 squadron at Northolt and it had risen to a fully fledged establishment incorporating the cream of Fighter Command.

The Next Generation of Jets 11

Following the final Meteor developments, two new operational requirements were formulated; the F3-48 and the F4-48. The former was a high subsonic single-seater fighter for daytime operations, which was awarded to Hawkers; the outcome of which was the Hunter which went into service with the RAF in large numbers and was most successful; it also sold well abroad on the export market. It was fitted with the Avon engine at first, then a version came out with the Sapphire and the final development was with the Avon RA-14. For a short while it held the world's air speed record, previously held by the Meteor at 623 mph. Neville Duke flew it and using a specially prepared engine with afterburner attained a speed of just over 727 mph. This record was soon beaten by the Supermarine Swift flown by Mike Lithgow using a similar engine, raising the record to 737 mph. Apart from the honour of holding this record there was a commercial motive too: as yet the Air Ministry had not decided which aeroplane should be ordered in quantity.

The Americans soon entered into the record breaking competition and in the hands of J. Verdin flying a Douglas A-4-D put the speed up to 752.9 mph. This was then beaten by Col Everest in a North American F-100-A at 755 mph soon to be followed by Haines in a F-100-C at 822.27 mph which for the first time just exceeded the speed of sound. Some months later Peter Twiss, a test pilot at Faireys, who had shared the development flying with Gordon Slade, the Chief Test Pilot, handsomely put the record up to 1,132 mph in the Fairey Delta way up in the supersonic range and there it stood for some time. This aircraft embodied an interesting new feature, a droop snoot nose. The nose piece was lowered for take off and landing to enable the pilot to see forward; this was necessary as the angle of the aircraft to the ground, due to the delta-wing form, blanked off the view. This feature has been adopted in the Concorde.

The Fairey Delta 2 in which Peter Twiss put the world speed record up to 1,132 mph.

While the Hunter was being developed, the Air Ministry asked Supermarines to produce an interim fighter based on the Attacker as an insurance policy to be in service if required earlier than the Hunter which was being designed to meet a specific requirement. The RA-7 type Avon was also chosen for the Swift but using re-heat; this was a mistake as the fuel consumption was greatly increased, thus reducing the radius of action considerably. There was keen competition between the two firms and each aircraft had certain advantages over the other, both suffered from engine flame-out when firing the guns. After exhaustive trials at Boscombe Down and at the Central Fighter Establishment, the Hunter was selected as the primary day fighter and only two squadrons of Swifts went into service in Germany for low level reconnaissance.

The Air Ministry decided to develop the Hunter to be capable of supersonic flight so work proceeded on the P-1083 which was to be fitted with an RB-133 engine. Just before the aircraft was ready to fly however the Air Ministry decided to cancel it. Digger Kyle now Governor of Western Australia, but then the Director of Operational Requirements, called on Sydney Camm and told him of the decision. It was a bitter blow not easily accepted and for a long time after Camm used to tell his visitors what he thought about it; he used to point at a spot on the carpet in his office and say 'That's where he stood. He came here and told me they didn't want it anymore, they're crazy.' But what I think was more tragic was that Hawkers did not proceed with it themselves as a private venture. I visited Camm several times to try and persuade him to fit a re-heated advanced Avon engine which would have given it a top speed of around Mach 1.8. Another factor was that Rolls-Royce needed an aircraft to launch a re-heated engine and this would have been an ideal application. It was estimated that the cost to Hawkers would have been two million pounds, not a lot when it would have ensured the replacement of the Hunter world wide, a matter of several hundred aircraft. It would have influenced the future success of the Mirage which became the natural successor to the Hunter; this was a fantastic opportunity lost.

The Hawker Hunter F-6

e Sea Vixen with two Rolls-Royce Avon engines; it was a development of the abandoned DH-110.

The other requirement F4-48 was for an all-weather fighter; both De Havilland and Gloster submitted designs and built prototypes. The specification called for two engines, radar ranging and navigation systems and guided missiles. De Havilland chose Rolls-Royce Avon engines for their De Havilland 110 (DH-110) and Gloster being in the Hawker Siddley Group used the Armstrong Sapphire belonging to the Hawker Siddeley Group and the rival of the Avon; this was healthy and it stimulated competition. We were not to be put off by this and on a number of occasions I visited Gloster to discuss with George Carter and Eric Greenwood the possibility of fitting Avons. The Gloster Javelin was finally chosen by the RAF; the DH-110 unfortunately crashed at Farnborough during the air display killing John Derry and Richards and sadly a number of spectators. This set the development back and the RAF lost interest. Several years later after considerable modifications to make it suitable for naval operations, it went into service as the Sea Vixen, an all weather naval fighter.

This period was one of gaining experience in flight at high subsonic speeds and in the behaviour of axial flow compressor jet engines. Both undertakings gave rise to serious problems; we in this country were not alone in trying to overcome them. I went over to the States to visit aircraft establishments and talk about the various difficulties being encountered. The effort being applied both at the factory flight test centres and particularly at Edwards Air Force Base was most impressive. The Americans were also actively engaged on supersonic aircraft and vertical take off both with propeller and jet, and with swing wings.

At Edwards I also saw large rocket engine test beds which were most impressive; I left there feeling that Britain was in quite a different league. They were very kind to me, Dick Horner who was in charge spent a whole day explaining everything (he later became Under Secretary of State for the USAF). Also the Chief Test Pilot of NASA, Scott Crossfield, showed me the Bell and

Douglas supersonic rocket aircraft. All these were on the farthest threshold of technology and way ahead of any other country at the time. I was very grateful to my old friend Roger Lewis at the Pentagon for having arranged it all. One of the most important aspects of my new job was looking for outlets for new and advanced engines. This meant the study of operational requirements, both in the United Kingdom and abroad.

The Rolls-Royce Avon had been selected for the Canberra, the Hunter, Swift, DH-110 (Sea Vixen), and the Supermarine Scimitar—all subsonic air-

A line up of Supermarine Swifts

craft. It had also been selected for the first of the V Bombers, the Valiant. Its main competitor at this time was the Armstrong Siddeley Sapphire which had been chosen for the Gloster Javelin, the Handley Page Victor, some of the Hunters and the advanced supersonic English Electric P1 (EE-P1) research aircraft, which later on when fitted with Avons became the 'Lightning'.

We were keen to become involved in supersonic aircraft; this meant an advance in engine technology, such as the development of re-heat or after-burning and cooling of components to cater for the rise in temperature through adiabatic heating at high forward speeds. There were two supersonic projects being built in Britain; one of them the Fairey Delta which was to have the Avon installed, the other the EE-P1 which was to have the Sapphire.

The Fairey aeroplane was most successful; much data was obtained from its flight tests culminating in gaining the world's air speed record. After the record was obtained the aircraft went to France where it concluded its low altitude high speed work; this was necessary as the regulations in this country precluded them being carried out here on account of sonic booms. The Dassault Company were helpful and gave assistance; they were most impressed by the design and performance of this aircraft and used it as the basis of their new Mirage design, which has since become world famous. The EE-P1 flew with the Sapphire as a research aircraft. The Air Ministry decided to develop it into a fighter and wrote a requirement for it. We then thought that we should try and persuade both English Electric and the Air Ministry to change the engine to an Avon.

After discussion with Elliot, Lombard and Lovesey, it was felt that the Avon using the new air-cooled turbine blades allowing a higher flame temperature, steel compressor casing to cope with the higher air temperature of supersonic speeds and the latest reheat system developed on the Fairey Delta, would offer a considerable advantage over the Sapphire. After several visits to discuss it with Freddie Page the Chief Engineer, he agreed to produce a brochure for submission to the Air Ministry and Ministry of Supply. We had meetings with Group Captain Peter Broad the Deputy Director of Operational Requirements, who accepted our proposal with much enthusiasm. The happy result was that several prototypes using the Avon were ordered and after successful trials which were carried out by Wing Commander Roly Beaumont, an old fishing friend of mine, it went into production and became the first supersonic fighter in the Royal Air Force. The most lucrative market was likely to be a common need in NATO to replace the obsolescent American types provided for the European countries under 'lend lease' such as the Republic F-84 and North American F-100.

The Germans were the first to begin a search for a new fighter; the numbers needed were substantial and so every manufacturer made a big effort to fulfil the requirement. Other countries including Holland, Belgium and Italy were likely to follow on, so whichever company or country won the contract they would be richly rewarded.

Britain was offering the Lightning and the Saunders Roe-177 (Saro-177); France the Mirage, the Swedes the Draken, and America the Lockheed F-104, the Chance Vought Crusaders, the Grumman F-11F Tiger and the Republic F-105 Thunderchief. The engine companies who were competing were of various nationalities; Britain was represented by Rolls-Royce with the Avon and Conway, De Havilland with the Gyron junior and Bristol with the

The Lightning with Rolls-Royce Avon engines, developed from the EE-P1 research aircraft.

Olympus. The USA was represented by Pratt and Whitney with the J-57 and J-75, General Electric with the J-79 and Curtiss Wright with the J-65 which was a Sapphire built under licence. France offered the SNECMA Atar.

Competition was all the keener because there were so many work hungry firms striving for survival and there was only this one large order to go for. The choice of aircraft and engine would likely spread to many other countries not yet in the market; thus the scene was set for a 'Battle Royal'. The situation was a salesman's paradise. The customer was lacking in up-to-date thinking and experience; the Luftwaffe having been disbanded after the war, with most of its experienced airmen now in civilian life. NATO planning was in its early stages so that there was no co-ordinated requirement; each country was able to choose its own equipment often dictated by national and commercial interest. The selling countries were motivated not by fulfilling a pure requirement to combat a genuine threat but by the desire to gain a commercial foothold in Europe. This introduced a control over what each firm was able to offer and how the proposal suited the political environment.

The result of all these factors was much confusion, skull-duggery and fairly amateur espionage!

Many of the firm's salesmen congregated in the Konigshoff Hotel in Bonn, sometimes it was known as the 'Lockheedshoff'! It reminded me of Lisbon during the war where all the spies were situated. To digress for a moment, I recall when spending a few days there in transit, I heard of a consignment of Swiss watches, coming to the airport from Geneva in a German Lufthansa aeroplane, which were consigned to the British Air Ministry in London. When they were reloaded on to a BOAC Dakota for onward passage to London, the German spy network heard of this and then informed the Luftwaffe so they could try and intercept and shoot it down!

The atmosphere in Bonn was rather similar! Competition between individual national companies was complicated by international and political suspicion. One American compiled a brochure to decry the merits of one of its brother competitors by publishing photos of the pilots killed flying it and describing how the bereaved families were suffering both financially and emotionally, going on to forecast that if that particular aeroplane was selected, many more fatalities would result with all the attendant misery.

We in Rolls-Royce were trying to offer our engines in all sensible and suitable aircraft, and these included in Britain, the Avon in the Lightning. This was quite a straightforward project as it was scheduled for the Royal Air Force.

The Germans were beginning to show interest in the SARO-177 which was a most interesting project as it combined a jet engine with a rocket engine; the intention was to cruise on the jet engine and use the rocket engine for short periods in combat. In Britain, the project was under reconsideration. The Royal Navy and the RAF had ordered five prototypes, fully intending to order the aircraft as an interceptor, when the Duncan Sandys White Paper on Defence had been published in 1957 which advocated the use of guided missiles

Previous page: The Supermarine N-113 Scimitar twin jet naval fighter

The SARO-177 which was never put into production.

in place of aircraft. This meant that certain aircraft projects were to be cut back and this included the SARO-177. This aircraft was designed to use the De Havilland Gyron Junior engine of 8,000 lb thrust. I had been a frequent visitor to the factory in the Isle of Wight trying to persuade Maurice Brennan the Chief Engineer and Captain Clarke the Managing Director to use a re-heated Avon instead. This would have improved the performance considerably as it would have enabled the aeroplane to fly supersonically without the assistance of a rocket motor. Derby however had not been enthusiastic; Lombard was not keen on mixed powerplants.

Help eventually came from an unexpected quarter: the Germans having expressed interest in the project were told about the possibility of fitting the Avon. Realising that the Avon would provide a supersonic performance, the scales were tipped in its favour and one was fitted and was finally selected in place of the Gyron Junior. The final outcome however was disappointing; just as the Germans were compiling a short list of selected projects the British Aviation Minister of Supply went over to Bonn to tell them that the British Government was about to cancel its contract for the SARO-177, but would not do so if the Germans would order it! What a way to sell an aeroplane! Naturally the response was 'Why should we have it if you haven't faith in it yourselves?'—So that was that.

I had been over in the States looking for aircraft that could be with advantage re-engined with the Avon and Conway and which were likely to have a successful future operationally. I had been to Chance-Vought at Dallas, Lockheeds, Republic, Convair and Grumman. The Chance-Vought F-8U2 Crusader already in service with the US Navy would accept the Avon, and a new project the F-8U3 would accept the Conway. They were however not supported by the US Government for sale to Germany. The Royal Navy however were very interested in the F-8U3 as a replacement for the Sea Vixen and Scimitar as a Naval Fighter. I tried to get both Faireys and Shorts to negotiate a licence to build it using a Conway. This came to nothing due to the effect of the 1957 White Paper. I got to know Paul Thayer at this time: he had recently joined Chance-Vought as a test pilot, having retired from the US Navy at Patuxent river; he had heard that there was a vacancy for a test pilot and so he sat on the doorstep until they accepted him, jobs not being easy to come by in those days. He has since become a legend and an example of determination leading to success; at the time of writing he is President and Chairman of the Board of the whole LTV complex. Mr McCarthy, the then chairman, was most kind and put me to work with Paul and Lyman Joseph to rough out the engine installation. They have been one of my favourite firms ever since.

The Lockheed F-104 had supporters at Derby as a candidate for the Avon; some felt that being a very advanced design it would be a way of gaining valuable experience; it was also on the short list for selection by Germany. I must say I was opposed to this as it had an appalling safety record on its initial test programme, also the USAF had decided not to adopt it, the J-79 was a difficult engine to replace and there were more suitable aircraft such as the Grumman F-11F Tiger on which to expend ones efforts. I did visit Lockheeds but found them unenthusiastic which pleased me. I couldn't help feeling that one only must be involved with aircraft that were likely to be popular with the pilots and had safe flying characteristics. We had had an experience during the war in being involved with incompatible installations, such as the Beaufighter, Barracuda and the early Halifax and Manchester and one had developed a 'nose' for avoiding them. They only brought trouble and gave the company a bad reputation. There is surely a knack in picking winners or at least avoiding losers.

The Grumman Tiger was a splendid aeroplane and a worthy partner for the Avon. It was a forgiving aeroplane and did not have vices; it had a good service record in the US Navy and with the Avon fitted would have a fine performance. I worked with Corky Myer, the Chief Test Pilot, who was in charge of selling it to the Germans and met Joe Gavin (now President of Grummans), the Project Engineer who had produced the aircraft. They agreed that an Avon installation should be studied and offered to the Germans, who would have been well advised to have bought it, but no, they were much too deeply impressed by the selling methods of Lockheed with the F-104. Lockheeds were really professional and were heavily supported by the US Government. The sale went through in spite of the parent country having discarded it themselves and its poor safety record. Most, if not all, of the other competing companies were amateur in salesmanship compared with Lockheed.

We also worked closely with Republic to fit a Conway in the F-105 Thunderchief which had recently been accepted into the USAF. Considerable time was spent with Murray Berkow and Gene Murphy in Paris persuading

The Hurricane, Spitfire, Gloster Meteor, Hunter, Javelin and Lightning

them to offer the aeroplane with the Conway instead of the American Pratt and Whitney J-75 or the British Olympus. We dreamt up a scheme whereby Rolls-Royce and Sud Aviation would co-operate with Republic to build the aircraft and engine in Europe for the benefit of the Germans. Considerable interest in the proposal was engendered; the Germans agreed to give serious consideration to it, provided a formal proposal was put to them. Murray and I decided we must get a document signed by all three principle Chief Executives to be submitted before the end of July. The problem was how to get the signatures all on the same piece of paper within the time allowed when the Chiefs were all in different places many miles apart and constantly travelling! Murray got Munday Peale his President to sign; I found Pearson at his holiday home in Wales, but we could not tie down Monsieur Herriel the President of Sud, and so it all came to nought.

Our last and final card to play was the French Dassault Mirage. The prototype using the French Atar engine had only been flying for a few months; nevertheless it was a worthy contestant for selection by the Germans. We approached Dassault and offered the Avon which gave a much better performance than the Atar engine. They were only half hearted at offering it, in fact they told us not to discuss it with the Germans as it was not their policy to displace the Atar. They hoped to obtain the order with a 100 per cent French product. Had they been willing to use the Avon with the reputation of Rolls-Royce behind it, they might have fared better. I will go into greater detail on the Mirage with the Avon engine in the next chapter, as this story is by no means over yet.

So the Lockheed F-104, known as the Starfighter was finally selected by the Germans; the Dutch and Belgians followed, then the Italians, Norway and Denmark. A large European manufacturing programme was formulated and so America and Lockheed achieved a highly successful sale. The unfortunate customers suffered however, and so many planes crashed that it became known as the 'widowmaker'. The West Germans were particularly badly affected and when no less than 178 of their Starfighters crashed, they sued Lockheed for three million marks compensation for the widows and dependants.

Lockheeds were overwhelmingly the most enterprising firm as far as selling was concerned; they were absolute professionals sparing neither effort nor expense in providing information at short notice in whatever foreign language was required. They also had the full support of the US Government. We learned that the Ambassador to Belgium, one of the customer countries, was an ex-director of Lockheeds! Looking back history shows that the choice of the F-104 was unfortunate; I believe that if Britain had really tried or if Grumman had had the support which Lockheed had, or the French been less selfish and more realistic, the outcome might well have been different and the countries involved would have been better served. So ended the struggle for the European or NATO fighter of 1957–8.

Perhaps the best summary of this struggle comes from an article in the September issue of RUSI by Bill Grunston: it was headed 'Lockheed Sweeps the Board'.

> 'Students of air affairs will be familiar with the incredible tale that unfolded in 1957–61 when many of the NATO air forces on the European continent shopped for their first super-sonic fighter. The behaviour of most was almost unbelievable and two countries that already had supersonic fighters of their own, Britain and France, went to extraordinary lengths to avoid any sensible or reasoned negotiations; Britain even cancelling the aircraft that had been judged first choice by Federal Germany, by far the biggest single customer. In the end of course the whole deal went to Lockheed for an aircraft which did not exist in the form required and which was of no interest to the USAF.'

It was at this time that I became Military Aviation Adviser to the Company, relinquishing the job of Export Sales Manager. It had been Lord Hives' dictum that sales and operational requirements didn't mix. He used to say that the operator or customer would not confide in you if he thought you were trying to sell him something. I also gained an assistant to help me, for on the European front, defence alliances were forming: first Benelux and then NATO, and this meant that there was soon going to be too much going on for one person in London to keep up a close enough liaison with each firm and all the various military requirements which were being put out for study. The Paris and Amsterdam offices were able to deal with indigenous projects but it became necessary to co-relate them to come into line with the growing European interdependence. Pearson suggested I should find someone to help me. I was fortunate after some searching to learn that Group Captain Mike Stephens was thinking of retiring from the RAF and his experience was just what was wanted. He, after a distinguished operational career in fighters during the war, had been on the operational requirements staff at the Air Ministry and was now in SHAPE on the same line of business. This meant that he was dealing with European military requirements and so he was just the chap. He joined us and helped me to integrate the European requirements. Expansion was rapid; America, Sweden, Germany and Italy and Australia all had to be studied and looked after and so we were kept very busy. The sales battle for Europe was virtually over once the Germans had decided to purchase and build the Lockheed F-104. There were however still as yet some uncommitted countries and there was still a chance of persuading them to buy the Mirage.

I went over to Paris with Sir Ralph Cochrane who was in charge of the Rolls-Royce Advanced Project Group to listen to a presentation on the Mirage given by General Pierre Gallois, Monsieurs Dassault's Military Adviser. You may recall that the Mirage was conceived after Dassault had studied the Fairey Delta which held the world's speed record. I was much impressed by the concept of the Mirage, its general philosophy and the plans for future development. I felt that this surely must be the vehicle with which Rolls-Royce should be associated with the Avon installed, not only to offer sales competition to the Lockheed F-104 worldwide but also to go hand in hand on future engine development in supersonic flying.

I went over to Paris to see Henri Deplante, Dassault's Chief Engineer whom I had known when he was with Sud Oust. He was a remarkable and likeable man; during the war he escaped from France and came to England where he fought with distinction. He was well known as a brilliant engineer. I explained to him what the Avon had to offer and its technical advantages over the Atar, the improved aircraft performance which would accrue, and made a plea that he should do an installation study. Massey-Cross, our Paris representative, was with me and when necessary he acted as interpreter and adviser on the diplomatic approach to the French. Henri was not very enthusiastic: the Dassault policy, partly dictated by the French government's attitude was that their aeroplane had adequate performance with the Atar and that they would prefer to sell an all French aeroplane including radar and weapon system.

I had recently been out to Australia to assess the fighter replacement situation there. I learned that they were within a few months of deciding to adopt the Lockheed F-104! However their over-riding requirement was to obtain an aircraft which had all the attributes of the F-104 but which at the same time would be able to go from Darwin to Singapore without refuelling, a distance of 2,000 miles. The F-104 could not do this, neither could the Atar Mirage. We realised however that the Mirage with the Avon would be able to do this as the Avon had a 15 per cent better fuel consumption and also had more thrust for take-off. We sowed this seed in the minds of the Australian Air Staff saying we would return in a few months time in order to present the Avon Mirage before Christmas 1959 which was the deadline.

I discussed this on my return with Tom Kerry who was on Lombard's engineering staff and was a past master at installing engines into different air-frames. He had always been most helpful and enthusiastic in doing preliminary schemes with me, ever since the Hucknall days when he had designed a jet trainer project for us to offer to the Dutch. Sometimes he had to do this in his own time; on several occasions he had been told by Lombard not to waste time on my hare-brained ideas, but to get on with designing power plants for American airliners! Even at this time the emphasis was more on the civil market than the military. Lombard usually came round in the end and once a project got going he would give it all the necessary support.

Tom Kerry drew an installation arrangement for a re-heated Avon in the Mirage which looked good. We went over once more to Paris to see Henri Deplante. He received us still without much enthusiasm. I was able to tell him this time of my visit to Australia and that the market there was now wide open if the Mirage could do 2,000 nautical miles ferry range which it could, if an Avon was fitted. Tom Kerry then produced his drawings to show that it was quite practicable. At first Henri thought the Avon too heavy, the air intake would be too small and so on. Then suddenly the penny dropped and he saw the opportunity being presented. Tom allayed his fears and criticisms, and soon Henri was saying 'This is good, no problem, no problem. When can I have an engine? We will produce a brochure, we must go to Australia.'

Having kindled Henri's interest, we had to get the support of the two managements and then find funds for a prototype. Pearson saw the possibilities and gave his support to the project. He came over to Paris where we had a meeting with Marcell Dassault and Henri. They agreed to share fifty-fifty in the cost of building an Avon prototype and also agreed that for the countries outside Europe including Switzerland the Avon version would be offered for sale. Inside Europe where Dassault were already offering the Atar version, Rolls-Royce would not compete. We invited Sir John Toothill (MD of Ferranti) to join our team and submit a proposal for the Ferranti Air Pass Radar System which was superior to the French Cyrano system. Thus we would have a larger UK content which should help the sale of the Avon Mirage in Australia.

A brochure was prepared and a comparison of performance between the Avon Mirage and the Lockheed F-104 was drawn up. Bernard Waquet, the Dassault sales manager who was on loan from the French Navy, and I set off once more for Australia. Since my previous visit, it had been necessary to appoint a representative in Canberra to look after our interests, anticipating a potential market in both military and civil aircraft. Who was better than my old friend Frank Carey, now a Group Captain and Air Adviser on the Embassy Staff. He had had a highly distinguished career in the RAF and had shot down as many enemy aircraft as anybody, quite a few being Japanese, in the Far Eastern Theatre, where the Australians were also prominent. He retired from the RAF and joined Rolls-Royce. Frank was a tower of strength and most helpful in the negotiations that were to follow. His help was invaluable also to Dassault who had no connections at all in Australia. Without the help of Rolls-Royce they would not have got to first base and the Lockheed F-104 would have been ordered.

We went the rounds of Government departments in Canberra presenting our project. This included meetings with the Defence Minister, the Minister for Air and Supply, the Chief of Air Staff Sir Frederic Scherger and we finally met the Prime Minister Mr Menzies at a reception at his residence. The result of all this was to defer the decision to purchase the F-104 until a team led by Scherger had visited Lockheeds, Dassault and Rolls-Royce to re-assess both aircraft. Meanwhile the Avon prototype had flown and was undergoing a complete programme of trials, showing great promise. The Australian teams finally discarded the F-104 and decided to order the Mirage. So far so good; we felt confident the Avon would be selected rather than the Atar, not only as its performance was better and its development potential greater, but also as we had a gentleman's agreement that the Avon version would be supported for the Commonwealth countries, as it had been our proposal to the French that the Avon Mirage should be offered to Australia to eliminate the F-104 Starfighter in the first place. Switzerland also was trying to make up its mind whether to buy the Swedish Draken or the Avon or Atar Mirage, and as they were quite likely to be influenced by Australia's choice the Australian deal became even more important.

I again visited Australia, accompanied by Ken Davies the Avon Project Engineer. We soon realised that the French, not content with defeating the Lockheed F-104, by comparing its performance with the Avon Mirage, were now offering the aeroplane with their own engine! This was on conflict with our unwritten gentleman's agreement. So there were now two brochures being offered to the Australians, the Avon version and the Atar. We were not allowed to see the Atar one! The makers of the Atar, the French SNECMA company had full access to the Avon performance and so knew what they must offer to try and equal us. This put us at a big disadvantage, as did the fact that they had the full backing of their government who absorbed any development costs; all the help we had was the loan of an engine for flight trials. The Avon flight trials had gone very well.

Realising that we were now up against severe competition, we offered to produce a larger diameter re-heat pipe, albeit at an increased cost. This would put the Avon performance well beyond the French. Unfortunately Rolls-Royce would not pay for this development, nor would the British government and so the Australians would have had to foot the bill if they wished to have it. The French, on the other hand, used all their political and commercial muscle and with their government's backing reduced the price of the Atar by £20,000. This coupled with a trade deal whereby the French bought wool and wine from Australia finally swung the balance in their favour. At one point I nearly discovered this commercial deal. I was having lunch with Monsieur Waquet and the French Commercial Attache; we were discussing Australian wine, I remarked 'Is it not true that the French buy Australian wine to blend with some of their Burgundy?' This was so near the truth that they vehemently denied it. I didn't realise the importance of this at the time; even if I had—I doubt whether anything could have been done to counter it.

The Swiss, as expected, then followed the lead of Australia, soon to be endorsed by South Africa; South America, and Pakistan, Israel and other countries soon followed. Another wonderful opportunity was lost by lack of enterprise and energetic support. Who was to blame? I put it down partly to excessive concentration on the civil market at the expense of the military and partly lack of support by the Government. We seemed to be battling our own 'built-in head wind'. The French are realistic and play to different rules. Another factor we failed to eliminate was the reluctance of the Australians to take a non-standard aeroplane; they felt they would be better served by having the same equipment as the French Air Force. They had in the past been using the Avon Sabre and before that the Nene Vampire, both aircraft had had certain problems which they put down to being non-standard and they did not wish to be out on a limb without the full backing of an Air Force behind them. How different from the Germans who took the F-104 when it was not even in use in the USAF! It must have been quite a close decision in the end; Air Marshal Scherger did tell me that if Rolls had agreed to fund the large re-heat pipe, it would have made all the difference.

Vertical and Short Take Off and Landing Development 12

Rolls-Royce was now becoming very active in promoting 'vertical take off and landing' (VTOL). Dr Griffiths the chief scientist realised that the thrust: weight ratio of a jet engine was becoming such that with further development it would soon be high enough (when installed in an aircraft) to give an overall value of greater than unity: thus enabling it to take off and land vertically. In order to test this in practice and develop a means of control and stabilisation a test rig known as the Flying Bedstead was built using two Nene engines. First of all it was tested tethered to the ground in case of non-function of the stabilisation system. Ronnie Shepherd carried out the testing, finally becoming completely airborne and flying slowly within a small area (approximately an acre) carrying out gentle manoeuvres under automatic control.

These tests proved the philosophy to be practical, although at the time it looked most odd and was excessively noisy! The preliminary results encouraged a number of designers both in the UK and abroad to build prototype aircraft to utilise the new lighter engine being built specially for this purpose by Rolls-Royce.

The Flying Bedstead with two Nene engines on its first test being flown by Ronnie Shepherd.

The three generations of Rolls-Royce lift jets: the RB-108, thrust/weight ratio 8.7:1; the RB-162-1, ratio 16:1, a third generation lift jet, ratio 20:1

The first of these light-weight lift engines was the RB-108 with a thrust: weight ratio of 8.7:1. These flew experimentally in the Short SC-1 and the Dassault Balzac. The next engine was the RB-162 which incorporated glass fibre for the compressor improving the thrust:weight ratio to 16:1. These flew satisfactorily in the Mirage 111-V. A final advanced engine, the XJ99, reached 24:1 but this has not yet found a home.

It was the practical power plant emanating from the basic Griffiths philosophy of direct lift by engine thrust; the intention being to use a battery or cluster of engines installed in the aeroplane, the number depending on the all-up weight of the aircraft. For example, if the take off weight of the aircraft was to be 30,000 lb, then eight engines would be required; the engine thrust would need to be 1.2 times the weight of the aeroplane to enable it to have a reasonable vertical acceleration. The thrust of the RB-162 was 4,700 lb.

This engine was being offered to the aircraft manufacturers, both in Great Britain and on the continent, but not to the US at this time; the reason being that we did not wish them to copy it and catch up on technology. The Bristol company were also interested in VTO, but favoured the vectored thrust method. This was achieved by having a large fan engine with rotating nozzles to give downward thrust. Two sets of nozzles were used, one set at the front deflecting air from the low pressure compressor and another set at the rear deflecting the hot gases from the turbine.

There became keen competition between the two companies as to which was the better way of achieving VTO and both Rolls-Royce and Bristol vied with each other in trying to persuade the various aircraft designers to use their own particular method. Adrian Lombard was the champion of the Rolls-Royce philosophy and Stanley Hooker lead the onslaught for the opposition. As usual money was very tight for research and development on engines and at this time there was no active operational requirement for a vertical lift aeroplane. The problem where to find the money? There was in NATO headquarters in Paris an office directed by the US Government called Mutual Weapons Development Programme (MWDP). It was administered by Col Chapman and its object was to select interesting projects that were worthwhile developing and which needed funds; the US Government would then support the project and at the same time take over certain rights in the design and production. The USA thereby bought its way into European technology and at the same time would advance the equipment of NATO.

This seemed to be a method of finding money to carry on while the British Ministry was making up its mind what to do. Unfortunately for us, however, Stanley Hooker with his silver tongue beat us to it and persuaded Col Chapman to support the Bristol method of vectoring the thrust and so his project gained the much needed financial support. Nothing daunted, Lombard persuaded Sud Aviation to use his method and they designed a very neat fighter using six RB-162 engines. While all this had been going on, there had been some research flying taking place in the UK on the Short SG1 flown by Tom Brook-Smith. This aircraft was fitted with four RB-108 engines of 2,000 lb thrust for lift and one for propulsion. This engine was the first light-weight engine designed by Rolls-Royce for vertical lift. Its power to weight ratio of 8.7:1 was not sufficient for it to be a practical power plant for a production aircraft and was being used solely for research into VTO. Brookie demonstrated that the multi-engined method of VTO was quite practical. He used to demonstrate their aircraft at the Farnborough show and in Paris and this greatly impressed the French that this was the best method of achieving VTO. The RB-162 was the next stage and a worthwhile advance and suitable for production. The Sud project with the RB-162 was still only a project and not funded; the immediate need was therefore to find funds so that progress could be made.

Lombard and I went down to Kingston to see Sir Sidney Camm and try and persuade him to use Rolls-Royce lift engines in his new project, the P-1127, which was being designed round the Bristol Pegasus. This we knew was not going to be easy as he had been working very closely with Stanley Hooker; in fact the P-1127 and the Pegasus had been designed around each other. However it was worth a try. Lombard went through all the advantages he could offer such as multi-engined safety, higher forward speed, better fuel consumption, etc., but Camm was not impressed. The Ministry was taking an active interest in the P-1127, partly because the American MWDP had agreed to put money into it and partly because the RAF was showing interest too. It seemed that we were going to loose out if we were not careful or at least we would have to look towards the French.

The thought occurred to me 'Why go to all the expense of building a complete new aeroplane which would only be marginally better than the Hunter in performance except for the advantage of VTOL?' Why not see if the

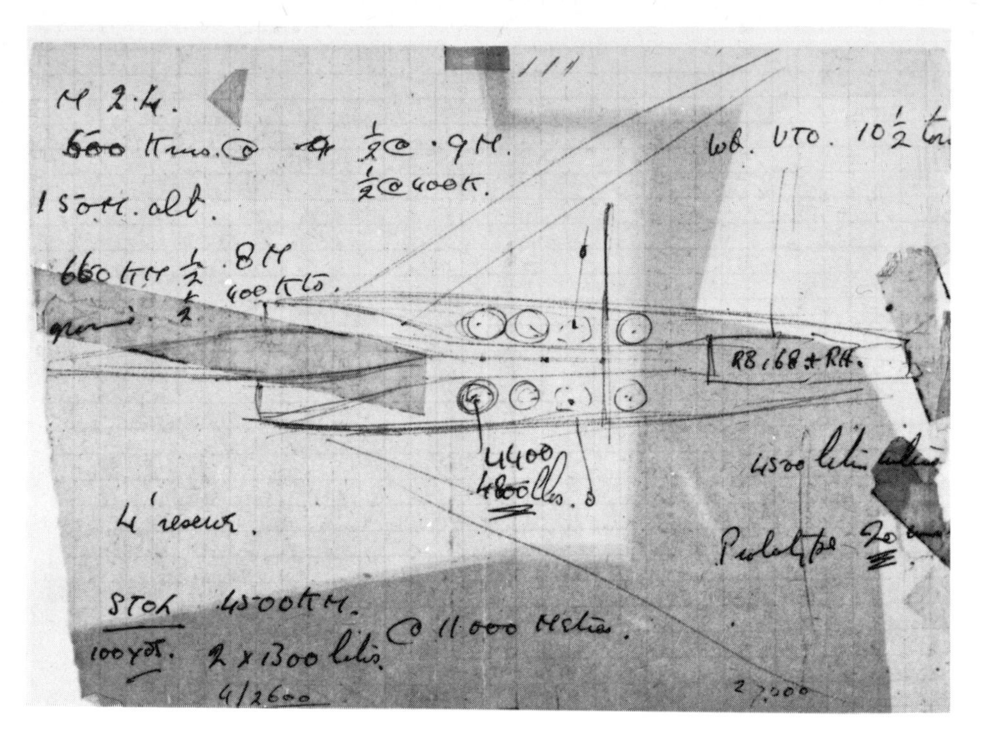

The original sketches made at a meeting between the author and Deplante, from which the Balzac and Mirage III-V emanated.

Mirage could be adapted to Short Take Off and Landing (STOL) rather than going to VTOL. STOL would provide a big operational advantage over the existing aircraft; the Mirage was a Mach 2 aeroplane; surely this if it could be done would be a good compromise!

We had on the drawing board an advanced small supersonic engine called the RB-153 which was being funded by the Germans for an experimental fighter they hoped to build. This gave 3,500 lb of thrust with re-heat. I sketched out a scheme for fitting two of these mounted under the wings of the Mirage on two of the four bomb pylons. They were able to swivel and point downwards at forty-five degrees for take off and landing. The Mirage would then have the ability to take off in a very short space as the total thrust would now be greater than the take off weight of the aircraft; it would have a speed greater than its present Mach 2; its range would be little impaired and it would have multi-engined safety in the event of a main engine failure. It would however loose the use of two of the bomb pylons but this could fairly easily be put right.

Before Pearson would agree to me going over to Dassault to offer it to them I had to convince him and Rubbra and Lombard that it was practical. This I managed to do in his office at Derby and so off I went to Paris to visit Dassault. I knew them all fairly well having had many meetings with them over the unsuccessful Avon Mirage and had been to Australia several times with them too. Firstly, I enlisted the help of Bernard Waquet, who had been helpful before, and he saw the possibilities of the project; then we discussed it with Henri Deplante and he too was willing to do a preliminary design. A few weeks later a meeting was arranged with Monsieur Dassault himself, Monsieur Valliere, Monsieur Cabriere, Henri Deplante, Waquet and the Rolls-Royce representative Massey-Cross as interpreter and myself.

This meeting has gone down in history at least so far as some of us are concerned. We hoped to persuade Dassault to put money into the plan and build a prototype and this was the big opportunity. I had also discussed with Henri the adaptation of a Mirage with eight RB-162 lift engines, but as a later project and one which might stand a better chance than the Sud aircraft with six engines which was a completely new project. I was very dedicated to Dassault and thought they were about the best people in the business. Henri had a preliminary drawing of the adapted Mirage which was also to be discussed.

After we had all shaken hands and the meeting started; Henri outlined the plan to Dassault; I could not follow it all in French but I got the feeling that he was not impressed and the various people round the table were looking uncomfortable. Dassault asked me through the interpreter why I thought it was a good idea and did I think the RAF would be interested? I spoke slowly in English and explained why it seemed to me to be a good compromise; it would make his already good aeroplane into something unbeatable until a fully developed VSTOL aircraft with eight lift engines could be produced and that it would solve for the time being the problem in NATO of getting away from long runways which were already targetted by Soviet missiles. He replied that he already had a very beautiful good aeroplane and he certainly wasn't going to spoil it by hanging engines under the wings and who was I to redesign it for him? He had a very good design team and anyway if he adopted this scheme people would think the present take off run was poor and needed all this extra power to take off! Innocently I replied that I did not agree with him, that he did not see the point, and I went on to explain how it would give him an advantage over the P-1127 and other competitors. By this time it was clear even to me that the atmosphere was becoming electric. He then wrapped his scarf round his neck (he habitually wore a scarf and kept his hat on) and started to get up, saying that he would not discuss the matter further and that if he was going to do VTOL he would do it properly from the word go.

The meeting broke up; the members of the Dassault staff looked distressed; particularly poor Waquet who was on loan to Dassault from the French Navy. He took me by the arm and said 'Let us go and have a cup of tea. I will tell you how my future is at stake.' We went along to where his speed boat was moored not far away and he took me for a bumpy ride on the Seine before going back to the club house for tea. He told me that he was to be sent back to the Navy, that I would never be allowed to visit Dassault again as M. Dassault was thoroughly displeased with the whole affair. The penny had dropped by this time and I suddenly realised the awe which everyone had for the 'Boss'; it was cushioned for me by not being able to follow the language very well. Nevertheless I did begin to feel a little worried on the flight home as I would have to report to Pearson and Lombard what had happened. It seemed to have been a pretty good failure!

This was not the end of the affair; how unpredictable the French can be! A fortnight later I received a message from M. Dassault asking me to go back to Paris to see him, I learnt that Lombard and David Huddie (The Managing Director) were going too. I knew that Lombard was not too keen on the lift engine version of the Mirage as he preferred the more elegant Sud scheme so I decided I would not attend the meeting. I was told afterwards that when Lombard and Huddie arrived, M. Dassault said 'I sent for Harker, he knows what the RAF wants; why isn't he here?' This rather naturally didn't please my bosses so things seemed to be going from bad to worse. However, they had a profitable talk and duly departed for Derby saying they would send me over within a few days time. When they got back they rang me up and duly pulled my leg unmercifully suggesting that perhaps I thought I was above mere mortals telling the best designers in both Rolls-Royce and Dassault how things should be done. I must say I think both Lomard and Huddle took it in very good part.

I duly paid a visit the following week and met with a friendly reception; Waquet had been reinstated and all was peaceful. They had decided to get straight on with the design of the Mirage 111-V with eight RB-162 engines and in the meantime and as a research project to fit the Balzac, which was a forerunner of the Mirage but subsonic, with six RB-108 engines. It was agreed with the French Air Ministry to join in on the funding of the RB-162 together with the Germans and so it became a tri-partite venture and development could now proceed parallel with the rival Bristol Pegasus and with Hawkers too as they now had a rival in the Mirage. So it wasn't too bad after all!

The Dassault Mirage III-V with eight RB-162 engines

Shortly after a serious requirement was written by the NATO Staff for a VTOL fighter called the NBMR-3. The French put forward the Mirage 111-V and the Sud project. Britain proposed the Hawker P-1154, which was a larger version of the P-1127 using a larger engine the Bristol BS-100 giving it a supersonic performance. There were other competitors too both German and American. Some months later the Balzac flew creating considerable interest in

The Dassault Balzac with six RB-108 engines

aviation circles: it was the next stage after the Short SC-1 in the development of jet lift. After some valuable development work was completed it unfortunately crashed. The full scale Mirage 111-V prototype was completed and started its test programme. It was most impressive, fulfilling what was expected of it: it took off and landed vertically and could fly at Mach 2. The main difficulty with VTOL aircraft was the auto stabilisation system. At sometime or other there have been failures in, I believe every system tested: the one in the Mirage was no exception and this aircraft eventually crashed also but not before having demonstrated that it was the only supersonic vertical take off aircraft to fly and fly well.

The NBMR-3 competition soon reached the stage when all the competitors had been eliminated except for the Mirage 111-V and the HSA P-1154. In order to make sure of getting the business for Rolls, we would have to get the power plant for the P-1154 in case the Mirage was not successful in winning the competition. We had been opposed to the single-engined approach to VTO, on the grounds of safety (and partly because it had been developed by Bristol!), and so had ardently been supporting the multi-engined Mirage. The Bristol BS-100 engine was not only a deadly rival for research funds and development, it was also highly favoured by the British Government and the RAF. Rolls at this time was short of development support, it seemed to be Government policy to build up Bristol and give them more work to do particularly on military projects. We were told that if the BS-100 engine went ahead, then Rolls-Royce would have to help but only as sub-contractors on production. This was anathema to me and I thought hard about how to get round this situation.

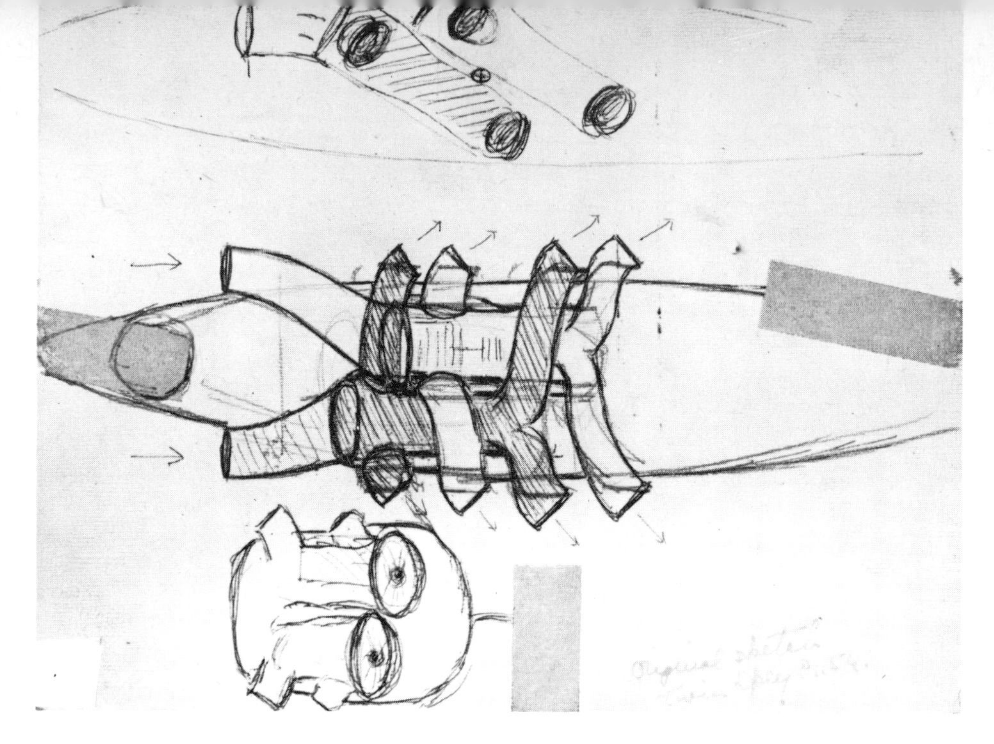

The author's original sketch for the Twin Spey P-1154

For some time now I had been looked upon by certain civil servants rather as a spanner in the works as I had been such an ardent supporter of the Mirage, while they had been dedicated to the P-1154. The situation became more acute when the Americans withdrew their support for the NBMR-3 competition on the grounds that the standard of technology necessary to produce a satisfactory VTO fighter had not yet been reached. The French decided to carry on alone with the Mirage 111-V and Great Britain wrote an operational requirement for the P-1154. It now became even more urgent, somehow to get engines into the P-1154, particularly as the French were going to use a SNECMA/Pratt and Whitley engine for propulsion in the Mirage and Rolls would only supply the RB-162 lift engines.

One day when discussing the subject with Air Commodore Ian Esplin who was Director of Operational Requirements at the Air Ministry, I learnt that he was worried about the cost of the BS-100 engine and the time scale in which it could be developed. This gave me a lead and opened up a possible opportunity of offering them an alternative. I went back to the office in Conduit Street and sketched out a scheme for putting two Spey engines into the aeroplane instead of one BS-100. I found that it seemed reasonable and worth pursuing provided that it would save money and could be done quickly; it also would have the added advantage of providing a margin of safety being a twin-engined installation.

I went up to Derby to discuss this project with Cyril Lovesey and Geoff Wilde and to see if they would set it out on the drawing board. They both saw the possibilities and agreed to give the matter serious consideration, after all if it was likely to prove successful it would put us firmly back into the military business. Adrian Lombard was away visiting Japan at the time, and being opposed to the vectored thrust solution for VTOL we thought it unlikely he would have agreed to the project department taking the job on as they were very busy on civil projects at this time. Once the scheme took shape the Boss, Sir Denning Pearson, was told about it and he gave it his blessing.

I then told Ian Esplin of the scheme saying that it would be less costly and available in a shorter time scale than the BS-100, and that it would probably have better performance and added safety as it was a twin. His reaction was very encouraging; he was apprehensive about the possible dangers of the single-engined BS-100; he welcomed our proposal and gave it his support.

The next step was to discuss it with Hawkers, so I wrote to Sir Sidney Camm outlining the proposal and asking him if he would like me to visit him and explain it more fully; he rang me up and asked me to Kingston to talk about it. He saw the logic of the proposal and asked how far we had got with it. When we told him, he said that he would like Lovesey to come down and tell him in detail how the engines might be installed and whether Rolls-Royce could develop a satisfactory plenum chamber burning (PCB) system and if a practical cross-over exhaust system could be engineered. Following this discussion with Camm and Lickley I was able to report with some confidence that the Spey project was receiving serious consideration, not only from Hawkers but the Air Ministry also, and that we should press on without delay with a design study. Lovesey and Wilde got busy and were able to produce a preliminary scheme on how the twin engines could be installed. The most difficult part was to arrange crossed over exhaust pipes, so that if one engine failed the other one would produce thrust on each side of the aeroplane and it would be in balance. This was not easy as each pipe was approximately 24″ in diameter and they had to cross each other neatly within the width of the fuselage and in a distance of some 48″.

We visited Camm, Lickley and their engineers and showed them the drawings. There was a great deal of scepticism at first followed by a sharp lecture from Camm on what he thought about Rolls-Royce wasting its time on lift engines and the Mirage instead of working more closely with himself. He felt that it was traditional for Hawkers and Rolls-Royce to work together, and pointed out the past successes: the Hart, Fury, Hurricane and the Hunter. He was very complimentary about Stanley Hooker and his colleagues, but thought it about time Rolls-Royce were again in the picture particularly as they were so good at doing things quickly. The cross over pipe scheme was severely criticised, but apart from this the other advantages were realised, although Fossard the chief designer remained critical in as much as he preferred the BS-100 single-engined scheme.

After a hard week's work by Geoff Wilde and his team we were able to return to Kingston with a half-scale model made in perspex of a most ingenious arrangement of cross over pipes which would enable the aeroplane to remain airborne and in balance in the event of an engine failure. Lombard returned from his trip to Japan, and happily he gave the project his blessing so that a full project study could be launched. We then went ahead and prepared drawings and a brochure and worked out a development and production programme together with an estimate of costs.

By this time the RAF, the Air Ministry and the Ministry of Technology were fully in the picture and the civil servants in London and the Gas Turbine Establishment at Farnboro started to evaluate in depth both the merits of the BS-100 and the Twin Spey installations; so now Bristol and Rolls-Royce were in full competition.

Official opinion was evenly divided and much lobbying went on between Rolls-Royce and Bristol in Government circles to gain support for their different power plants. The P-1154 was a joint Royal Navy and RAF project, having been promoted by the Ministry of Defence Central Staff and championed by Air Vice-Marshal Neil Wheeler, an officer who had much experience in the writing of operational requirements. It was hoped to save money by using the same aircraft for both Services as the Americans were trying to do with their F-111. The Navy were not too enthusiastic about the BS-100 engine on account of there being no margin of safety in the event of an engine failure and the lack of range. The Twin Speys on the other hand gave more take-off thrust and better fuel consumption and with a single-engine safety margin.

After several months of study the time came when a decision had to be made in order to meet the required time scale into service; it was a close thing and the decision was to retain the BS-100 engine installation. We were naturally disappointed having rocked the boat and come so close to displacing our rival. The Twin Spey layout was patented and brochures sent to American aircraft manufacturers for their consideration.

While this appreciation was going on, the Royal Navy were becoming less interested in the P-1154; the range and endurance were inadequate and in order to overcome this long range tanks were needed and as this increased the all-up weight it had to be catapulted from the deck and the bicycle undercarriage would not fit on the catapult. This led the Royal Navy to start looking around for a more suitable aircraft such as the McDonnell F-4 (Phantom). This aircraft was in use in the US Navy and in the US Air Force. The impact of this result was not lost upon Sidney Camm and his colleagues who realised that the Phantom was faster than the P-1154 and saw the advantage of again looking at the Twin Spey version. They now became more enthusiastic about it, as it could still be acceptable to the Royal Navy and might overcome many of their objections.

It is interesting to record that the top management of Rolls-Royce thought it imperative that the order should be placed to help the company out of a financial dilemma. The following is an extract from a letter from Sir Denning Pearson to Sir Arthur Vere Harvey who was chairman of the Conservative Defence and Aviation Committees:

'While we have been bending our energies to building up a civil export business, the military business has been going to Bristol Siddeley, with the result that today we can forsee no new major military project coming to Rolls-Royce, unless of course we can substantiate our claims for a Twin Spey to replace the BS-100 in the P-1154. Incidentally the same engine is suitable for the RAF tactical transport which is growing out of the OR-351 requirement.

'To try to summarise the position, we urgently need a government contribution on a 50/50 basis to the development of the R.Co. 43 Conway for the super VC-10 to relieve our immediate financial position and in the long term a major military project to redress the balance between our civil and military work, such as the Twin Spey in the P-1154 and tactical transport. If we can get neither of these then the future of Rolls-Royce as an independent firm is grim indeed.'

The Navy began quietly building up a case to buy the McDonnell Phantom from the States. A party from McDonnell consisting of Charles Forsyth, Admiral 'Cat' Brown and Mr Dickman called to see me at Conduit Street and said they believed there was an interest in the RN to purchase the Phantom instead of the P-1154; if this was truly so, they felt it would be a more attractive proposition if Spey engines could be fitted instead of the General Electric J-79, as they would provide more thrust for take-off and a better fuel consumption. Of course this suggestion fired my enthusiasm, for not only would the Spey installation make a fine aircraft better but it would be a spur to the chances of getting the Spey accepted for the P-1154. Maybe the Navy would choose the Phantom and so halve the number of P-1154s required, but there was still the RAF who might go for the Spey rather than the BS-100.

The situation was getting more complicated because we were also fighting a battle on the new transport requirement, which I will describe more fully later, suffice to say that the Rolls engine was also chosen for that too. We were now beginning to find ourselves in a rather good position; from a most depressing state of affairs only a year back we now seemed to have captured the market for new military aircraft. Oh what a chancy business this was! However, it didn't seem reasonable that Rolls-Royce should have all the business, partly on the grounds of capacity to do the job and partly to satisfy the policy of the Ministry which had been to equalise the amount of orders between the two firms, and so negotiations were begun to settle the division of work.

It is not absolutely clear what transpired between the top man in Hawker Siddeley (HSA) and his counterpart in Rolls-Royce but it is believed that they met to discuss how they should share the engine work. Rolls-Royce were in the position of winning not only the Spey for the Phantom for the RN and for the P-1154 for the RAF, but also the Medway (a larger version of the Spey) for the Armstrong Whitworth (AW) 681, (this company was also in the HSA group). If the Medway went ahead there was a good chance that it would also be selected by the Swedes for the Viggen. This Swedish fighter would use the

The Spey Phantom

Pratt and Whitney JT-8 engine, although the Medway was more suitable, simply because at this time the Medway had no other home, owing to the Ministry not funding further development on it.

Rolls-Royce were willing to relinquish the P-1154 provided they had the Medway in the AW-681 and this seems to have been what was arranged between them. It was shortly after this meeting that doubts crept into the minds of the Rolls-Royce engineers whether or not they could develop a satisfactory PCB that would light up quickly enough for landing and also whether it would give sufficient thrust augmentation. A deputation went down to Kingston to see Sir Sidney Camm to tell him about this. Rather naturally he was livid and said it was rather late in the day to tell him this now that the Spey was likely to be selected. The Deputy Chief of the Air Staff was also hopping mad and this is where I came in for some pretty harsh words. Now that the Spey had been withdrawn the BS-100 project was continued until the whole project was finally cancelled by the new labour government which came into power shortly afterwards.

The irony of all this was that while the industrial tycoons were deciding amongst themselves how to divide the spoils between the P-1154 and the AW-681 unbeknown to them, the RAF had already decided to scrap the AW-681 and buy the Lockheed C-130. The final outcome was that the Spey went into the Phantom for the Navy and the RAF, and later on the Navy air arm was abolished and the RAF took over their Phantom! Who could have guessed this remarkable and unpredictable outcome?

The story behind the Spey Phantom being accepted by the Navy was an interesting one in itself. When I told the powers that be up at Derby that the Navy were now becoming keen to adopt the Phantom and that the Spey was necessary for it to be suitable for Carrier operation, it evoked little interest at first. How could the RN get away with buying an American aeroplane when the British P-1154 was being developed for them? However, my friends in the design office sketched out an engine installation which showed that it was possible to accommodate the Spey. The next thing to do was to expose the top Rolls-Royce engineers to the Admirals who were of course full of enthusiasm so I arranged a dinner party at the Savoy with Admiral Sir Frank Hopkins, Admiral Johnnie Ievers, who had been Commander in Charge of flying on board *Ravager* when I went through the barrier on my first deck landing twenty years before, and Captain 'Winkle' Brown in charge of operational requirements. Sir Denning, Adrian Lombard, Cyril Lovesey and myself represented Rolls-Royce. After this dinner party it was realised that the Navy were very seriously looking at the possibility of buying the Phantom and having the Spey installed if the P-1154 did not come up to expectations. There were many problems to be overcome. First there was keen competition from General Electric in USA, who did not want their J-79 engine to be displaced by the Spey, then there was the rising cost of engine development, the reluctance of the RAF to have the Spey, as the superior performance of the Spey was not essential for them and they could buy a greater number of standard Phantoms as they were two-thirds the price, and finally the full estimated aircraft performance was not achieved, and the cost was too high to enable this version to compete with the standard one in the export markets of the world.

Studies of the Spey installation in the Phantom now went ahead at high priority; technical brochures and cost figures were produced and a general sales case was made out describing all the advantages that this engine would give compared with the General Electric engine. The Navy were satisfied at first on the grounds of performance, provided the engine handling would be as good as on the J-79 and that the cost would be reasonable. The RAF on the other hand, who by now were realising that the P-1154 might be cancelled and that they too would require the Phantom, were highly critical of the cost.

General Electric saw an opportunity of counter attacking and so promised elaborate modifications which would increase the thrust of their engine, hoping that it would be acceptable for carrier operations on the smaller British Carriers. The decision was in the balance for some time; the Ministry of Technology favoured the Spey mainly on the grounds of supporting a British engine and not spending more dollars than they need. It was accepted that the performance of the Spey was superior, but everyone was still worried about the extra cost.

While the final decision was still in abeyance a sudden crisis arose; Sir Denning Pearson and Sir David Huddie were in California talking to Lockheed about the Tri-Star and the RB-211. I was having one of my routine lunches with Admiral Hopkins who told me that Dennis Healey the Minister of Defence had just been told by the Ministry of Technology that the price of the Spey would be £157,000, when previously he had been told by Rolls-Royce that the price would be fixed at £137,000. This meant £20,000 extra for each aeroplane; this would be too much and so to keep within the Defence Budget they would have to buy the standard Phantom which was much cheaper.

I telephoned Adrian Lombard to tell him of this serious turn of events, he was just on his way to London and came to see me in my house late evening. We put a call through to California and spoke to Pearson and Huddie; Pearson decided to fly straight back to London and asked me to fix an appointment with the Minister for the next day which I did. He explained to the Minister that the price given by Rolls-Royce still held good and that the extra £20,000 was added on by the Ministry of Technology as a contingency. The outcome was that the cost of the engine had to be reduced and this was done by a sacrifice of performance at high Mach numbers by specifying an inferior metal in the turbine blades. It was a near thing and we just scraped through. Captain Hickson, who was the project officer at the Ministry of Technology, had been most helpful all along and I think we owed much to him for the successful adoption by the services of the Spey installation.

The contract was signed and the engineering went ahead, the engines were delivered on time to McDonnells at Saint Louis and flight trials got under way; they were promising at first and then troubles in the reheat system were manifest. It was not realised at the time how serious this was until Air Vice-Marshal Derek Hodgkinson (ACASOR) and A/C Colin Coulthard (DOR) flew the aeroplane at Saint Louis.

I had invited them to the Air League Ball to join the Rolls-Royce party with Sir David Huddie as the host. I met them for a drink beforehand and found they had just returned from the States that afternoon; they gave it to me straight that they considered the Spey Phantom non-operational until the engine and reheat handling had been vastly improved; this hot news almost put a damper on the party but the next day, when Huddie arrived back in Derby, urgent moves were put into motion to have the prototype sent over to Hucknall for Rolls-Royce to investigate the problem on the spot rather than 5,000 miles away. The problems were duly solved and the aircraft entered service in both the RN and RAF.

It seems that the development of aircraft to meet requirements never runs smoothly. Either economic or political pressures intervene or else the requirement changes, or unforseen delays occur. It is a foolish management who gives up too soon, or is too rigid in its ideas or is too much influenced by the financial and profit motif. The success usually comes through relying on experience, listening seriously to the customer's opinion and pressing on without delay and with determination. Dassault have been shown to be the most successful post-war company and they have exhibited these qualities.

While all this had been going on, rumours of problems had got to the press and so considerable diplomacy had to be exercised by all concerned. Air Commodore Peter Brothers, an old friend of mine from war days, was the Director of Public Relations for the RAF. He bore the brunt of questioning by the press. It was necessary for him to satisfy the press but at the same time not cause alarm in the States where negotiations were at a delicate stage with the RB-211 in competition with General Electric for the Tristar and the DC-10 airliners. Had it got around that Rolls-Royce were behind-hand on the Spey it would have influenced the value and credibility of their promises on the RB-211. At the same time there was a keen competition going on in Germany on the MRCA for selection of engine. This was between Rolls-Royce and Pratt and Whitney for an advanced fighter engine; any doubts on the ability of Rolls-Royce to meet its promises on the Spey Phantom would have reacted unfavourably and given ammunition to the supporters in Germany of the Pratt and Whitney engine of which there were many in high places.

Chapman Pincher of the *Daily Express* and Teddy Donaldson of the *Telegraph* were made aware of the situation by Pete Brothers and they withheld any information which could have been damaging. I had several lunchtime chats with Chapman at this time and gained respect for his integrity. When things got difficult we could always talk fishing, a subject we both had in common. He is a prolific catcher of salmon and has access to many famous beats of the best rivers. Hindsight suggests that he had printed the situation in a damaging light which certain journalists of the anti-aerospace lobby might well have done, the RB-211 might not have been ordered for the Tristar, nor the RB-199 for the MRCA. It is interesting to speculate where the company might have been today!

These stories show what an uncertain business it is to try and follow up the right path, who can tell what the right path may be when pure customer requirements are bedevilled by commercial and political interest? What might have happened if Australia and the Swiss had elected the Avon in the Mirage? Probably there would not have been any Phantoms in the RAF! What might have occurred if the Spey Mirage IV had materialised instead of the GE F-111. Probably the AFVG would have gone ahead with the French instead of the MRCA with the Germans.

It is a chancy business at the best of times; if only Rolls-Royce and General Electric or Pratt and Whitney had got together to build an engine for the large wide bodied Civil Transport, instead of cutting each other's throats and going it alone; who knows where the company would have been today. This living dangerously was in ones blood, but the risks to the unknowing shareholders were great and not understood.

SLAM: The Flying Bomb 13

In the early 1960s discussions had taken place between America and Britain about the production of a nuclear deterrent. It had been decided that Skybolt, an air-launched ballistic missile, should be developed and produced. It would be carried by the American Strategic Air Command (SAC) and America had agreed to sell 100 of them to Britain as a means of extending the operational capability of the V Bomber Force into the next decade.

In 1962, however, the United States Defense Department decided to scrap the project as it was no longer considered to be cost effective, and at that time it had not had a successful flight. They decided instead, to rely upon the Polaris missile which would be fired from their nuclear submarines. This would be used in conjunction with Minuteman and Skybolt would be dispensed with. The deal was complicated by Britain's desire to obtain the maximum compensation from the United States for cancellation of Skybolt and also to maintain as much control as possible of the deterrent. The Americans did not wish to weaken the Anglo-American alliance, but did wish to dilute Britain's independent control. Thus the Polaris deal seemed to be the best compromise to the politicians.

President Kennedy and Harold MacMillan met in Nassau on 18 December 1962 to discuss the implications of this cancellation and agree on an alternative solution. Whilst talks were in progress, the US airforce achieved a satisfactory flight, but it was too late and the Prime Minister was told that their previous agreement regarding Skybolt was no longer valid and that he could have Polaris instead. This was a big blow to the RAF for it meant that the nuclear deterrent would be taken away from the RAF and given to the Royal Navy who would be allowed only to build four Polaris submarines, leaving only two operational at any one time! Had Skybolt been retained, the RAF could have deployed the whole V Bomber Force for many years.

Any discussion which affected the RAF, of course affected us, and I felt sure that British industry could come up with a solution to unfetter us from American ties and influence. It was near Christmas and over the holiday I could think of little else but how to avoid the decision to buy Polaris, being taken. After all, the RAF had been anticipating operating with Skybolt for many years; the V Bombers had been carrying Blue Steel, a rocket-propelled, short-range supersonic flying bomb since the early 1950s. I reasoned that

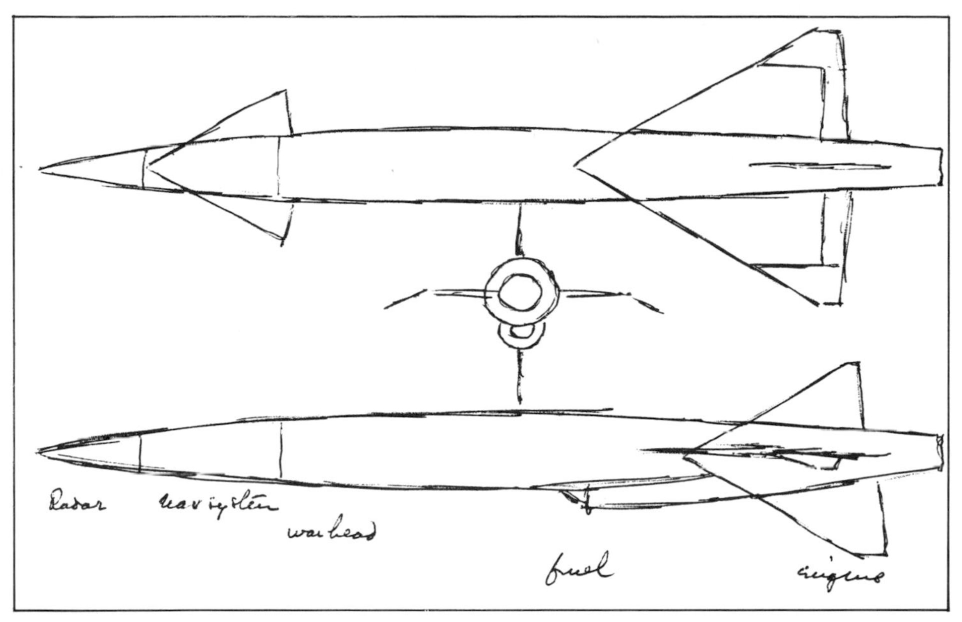

The SLAM flying bomb proposed by the author to try and avoid the nuclear deterrent being taken out of the hands of the RAF.

Britain should be able to produce her own flying bomb—why not use a Rolls-Royce jet engine instead of a rocket which could offer a supersonic capability and a range of some 1,000 miles, but at low level where it could not be intercepted. It all seemed to me to have been a dreadful political decision.

I discussed the project with Challier who thought it might be feasible. Encouraged, I then sought out John Keenan, who was the Chief Designer of Rolls-Royce's Forward Planning Department under Sir Ralph Cochrane, and asked him if he would do a preliminary design of a flying bomb using an RB-145 engine. If the project was feasible, I would then try and persuade one of Britain's aircraft companies to come in with us and jointly sell the idea to the Air Ministry and the RAF. The study drawn up gave these dimensions: the missile would be 35 feet long with a diameter of 3 feet and a wing span of 11 feet. Propelled by a Rolls-Royce RB-145 engine, using shelldyne fuel giving between 2,000 and 3,000 lb of thrust, the flying bomb would travel 1,000 miles at a speed of Mach 2 (that is twice the speed of sound), carrying the necessary 600 lb of nuclear war head. This was very promising provided we could find a guidance system to give the required accuracy.

The next step was to tell the top men at Rolls-Royce and get their blessing, so that design time could be put in on it. Sir Denning Pearson agreed that I could pursue the matter further, but nobody was really enthusiastic! I tried to interest BAC and HSA, but they were already in the missile business and had no use for it. I then approached Shorts in Belfast and talked to David Keith-Lucas the Chief Engineer, who became reasonably enthusiastic and agreed to do a study in some depth. Peter Hurn of Elliots agreed to co-operate on the guidance system and offered a gyro which was alleged to be accurate to within a mile after an hour's flight, and also to incorporate a terrain avoidance system. A joint brochure was produced, which stated that the project could be developed within four and a half years at the cost of some 18 million pounds. It could be carried on the V Bombers, the TSR-2, the VC-10 and even on high speed motor launches, etc. It would be dropped off at 35,000 feet, dive down to sea-level lighting up on the way down and then travel to its target at Mach 2 under inertial guidance, so avoiding high ground on the way; it could not be intercepted and, incidentally, it could do quite a lot of damage on the way by just flying at 100 feet at supersonic speed. It had the advantage also of allowing the V Bombers to remain at altitude; whereas in order to continue to be viable they would have to operate at low altitude thus seriously reducing their fatigue life.

Now we had to try and sell the idea to the Ministries! The RAF showed interest rather as one would have imagined, for if it was produced it would keep the deterrent in their hands. I enlisted all the support I could think of, including the press: David Devine wrote a piece in the *Sunday Times* and Derek Wood wrote it up in *Interavia* to help us get the French interested—perhaps it could be used on the Mirage 4. I discussed it with Ronnie Lees who was now Deputy Chief of the Air Staff, also with General Paul Stehlin, the Chief of the Air Staff in France; Ronnie gave warning that there was a political angle to this on account of the TSR-2 being used in the nuclear role, which would not be popular with the Socialist Opposition. This made matters rather sensitive as the TSR-2 was under attack at this time. I invited Denis Healey to lunch at Grosvenor House as he was the Shadow Minister of Defence to see what he might think of it in the event of Labour getting in at the next election. He would not commit himself, rather naturally, although one gained the impression that the TSR-2 would not be used as a nuclear deterrent. I then arranged to see the Air Minister, Sir Hugh Fraser, accompanied by Ronnie Lees. He was interested up to a point, as he wanted the RAF to continue as the custodian of the deterrent, but it was clear that nobody was going to stick their necks out to challenge the decision made at Nassau by the Prime Minister that the Polaris submarine should be the chosen instrument.

I still believe this 'flying bomb' using a jet engine would have been a very effective weapon and an economic method of maintaining the deterrent. The USAF had Hound Dog which had long range but was subsonic and therefore obsolescent. Ironically too, the Russians suffered no decision inhibitions. They adapted their Tupolev TU-20 (NATO coded Bear-B) subsonic four-turbo prop strategic bomber in the 1960s to carry an air-to-surface 'flying bomb' which had a range of only 400 miles. It was NATO coded as AS-3 Kangaroo. For years now, these Russian TU-20s of various types have been shadowing the coasts of Britain probing our electronic defences. This Bear B/Kangaroo combination had been revealed a year previously at the 1961 Moscow Fly Past in honour of the Soviet Aviation Day celebrations.

As I write this book some fifteen years later, having failed to persuade the powers that be to proceed with the project; I read in *Aviation Week* of

21 January 1974 that 'The USAF is pressing for the development of an advanced "long range nuclear missile" that would be launched from a military version of the Boeing 747 Jumbo Jet or the Lockheed C-5-A military transport'. The American Defense Secretary, Mr James Slesinger, is said to favour the air-launched missile despite its technical difficulties and the haunting precedent of the cancelled Skybolt. I also read in the *Daily Telegraph* December 1975, that the SALT (Strategic Arms Limitation Talks) talks between the USA and USSR which have been going on for several years are now held up as the Soviets object to the USA developing just such a weapon as SLAM, but called a 'Cruise missile'. This missile is similar in concept to SLAM, but comes fourteen years later and is alarming the Soviets! The *Telegraph* headlined the article 'US Flying Bomb Could Save Atom Holocaust' and explained that if a non-nuclear warhead were to be fitted it could strike deep into Russia hitting strategic targets with an accuracy of thirty feet and without the risk of triggering off nuclear retaliation. It went on to say:

> 'The cruise missile is a modern version of the Nazis' flying bomb! The US Army and Air Force are developing a long range version able to fly at low altitudes for about 2,000 miles and strike within 30 feet of the target! It is propelled by a turbo fan engine which gives it long range but only enables it to fly at subsonic and transonic speeds, whereas SLAM used a pure jet engine which gave it twice the speed, but half the range. However the use of a jet engine using shelldyne jellified fuel of high calorific value rather than a rocket was the real "break through" on propulsion. The accuracy of the Cruise Missile is much better than SLAM enabling it to carry high explosive, if required with effect, if nuclear is not deemed necessary. This of course is a normal development which has taken place in the intervening period.'

Mr Burt of the Institute of Strategic Studies says 'The "Cruise Missile" is the forerunner of a new generation of weapons technology that promises to alter how strategists think about deterrence and defence in the nuclear age.' The following quote from *Sunday Telegraph* 25 January 1976 sums up succinctly the impact, that the Cruise Missile is likely to have on the present and future arms talks.

> 'The Cruise Missile is astonishingly accurate and its versatility defies the normal categorisation of weapons. It can carry nuclear or conventional war heads and be launched from anything from a submarine to a jumbo jet. The most ominous race of all appears to be that between technology and diplomacy.'

'The real problem,' says Richard Burt of the Institute of Strategic Studies, 'is not what SALT will do to the Cruise Missile but what weaponry like the Cruise

Missile will ultimately do to SALT.' A remarkable state of affairs and a great opportunity which could have been achieved so much earlier and by Britain had been missed. Lord Hives once said 'It is no good having the right idea at the wrong time' and perhaps this was the problem of SLAM.

The conclusion one must draw from what proved to be a piece of political manipulation, albeit with hind sight, is that planning for defence at Service Staff level is only the beginning of long, complex negotiations which can be influenced by politics, economics, foreign policy and what actual equipment is required to defend the Country! The greatest political error in this context, of all time, it will be recalled was in 1946, when Sir Stafford Cripps, then President of the Board of Trade, was responsible for instructing Rolls-Royce to supply the Soviet Union with a small number of Nene and Derwent jet engines. This was just about as serious as Fuchs obtaining the secrets of the Atomic bomb. Even Stalin never believed he could obtain a British engine which he sorely needed to power the new Soviet aircraft designs: his own engines, lagging behind in technology were holding up his whole aircraft programme. When one of his ministers had suggested trying to obtain a British engine he is on record as saying 'What fool would sell up his secrets?' This dreadful interference by a British politician enabled the IL-20 (Soviet Canberra) and the Mig-15 fighter to take an active part in Korea against American and British fighters. Within six months of purchase virtually every major airforce design bureau in the Soviet Union had initiated design studies for combat aircraft utilising the British engines.

This monumental post-war political error was followed eleven years later by another and this time by a Tory Government. The notorious Duncan Sandys White Paper of 1957, by putting the emphasis on guided missiles instead of fighter aircraft, put British industry far behind other nations in the development of supersonic aircraft. Then came the third political interference when Britain capitulated to America by her obedient acceptance of President Kennedy's cancellation of Skybolt. The fourth one came about three years later in the spring of 1965 shortly after the Socialist Government came back into power lead by Harold Wilson.

Cancellations Galore 14

The change of Government in 1965 caused a complete upheaval in the aircraft industry. Within a very short space of time, all the major projects had been cancelled; these included TSR-2, the Hawker P-1154 and the Armstrong-Whitworth 681. Admittedly they had been started late and would hardly have gone into service before the existing aircraft, the Hunter, Argosy, Hastings and Javelin and the V Bomber force had become obsolete, but although the programme was an expensive one, the money involved would have been in sterling. The real reason for these cancellations has never been officially stated and probably won't be known until the Cabinet papers are made public. It is however almost universally believed that a deal was negotiated between the Labour Prime Minister and Minister of Defence with the President of USA and his Minister of Defense in which the slumping pound after the election was to be bolstered by the dollar. In exchange the competition of the British Aircraft Industry against that of the United States for world export markets would be eliminated. There must have been some restricting clauses otherwise the Spey Mirage would have stood a fairer chance of being accepted instead of the F-111. Stephen Hastings in his book *The Murder of the TSR-2* quotes:

> 'On December 6th and 7th the Prime Minister saw President Johnson in Washington at the height of the financial crisis. Backing had to be found for sterling. We know that the vast cost of the American aerospace programme was worrying the American Administration and Macnamara had resolved to lighten the load by selling American Defense equipment abroad. The success of the "hard sell" was already a byeword as witness the export of the Lockheed F-104 Starfighter to Germany and Holland and Belgium and Italy, a moderate aircraft for which the Germans had no easily recognisable requirement in the first place. By now a formidable team of salesmen had been assembled in the Pentagon under Henry J Kuss Junior.
>
> 'The proposition that America's Allies should cease to compete with her over military aircraft had been remorselessly plugged in Europe for some years. American resources were so vast that she was bound to win in the end, so why did we British not relax and come to an arrangement by which we should accept a proportion of the

total manufacture? I remember hearing it all in Paris from more than one plausible American salesman, both from the industry and the State Department as early as 1956. By now the Americans were so keen to break the British Aircraft Industry that Henry J Kuss and his colleagues actually offered the C-130 to the British Government at a figure between £700,000 and £800,000* while the unfortunate Australians had to pay £1,300,000 for the same aeroplane.

'Into this atmosphere of steely determination and super salesmanship walked a new British Prime Minister, with little business experience, with a committment to review all our important aircraft projects, with no manifest concern for British Aviation anyway, and with a begging bowl in his hand. It would have been surprising indeed if the Americans had failed to bring off a deal; but, to judge from some reports, even they did not expect the push over to be quite as simple or extensive as it turned out to be.'

With hind sight we were really bucking a head wind but at the time the options we were supporting at Rolls-Royce all seemed reasonable. The chief government planners and the Heads of industry and the Service Chiefs were as much in the dark concerning these political decisions as we were. It makes logical military requirements seem relatively unimportant when it can be seen that they can be set aside by politicians who may or may not have the real welfare of the country at heart. There is a great danger in peace time of insidious influence from a potential enemy gaining support for reducing armaments and defence budgets. When one compares the straight forward line of development to satisfy military requirements in war time with the tortuous machinations in peace time no wonder that the defences of the West are inadequate to restrain the single-minded intentions of the East. Any attempts to alter things were doomed to failure: it was all pre-arranged by the politicians but we did not know this at this time and so we tried to steer things our way.

The Hawker P-1154 was cancelled in favour of the Spey Phantom. At least Rolls-Royce did not suffer from this! Then the AW-681 requirement was cancelled. This had been under discussion for several years and finally Armstrong Whitworth had been selected to build this medium range V/STOL transport. Rolls-Royce had been in keen competition with Bristol Siddeley who were offering the Pegasus vectored thrust engines. Rolls-Royce were finally chosen using a combination of Medway engines with swivelling cascades to deflect the thrust and a battery of RB-162 engines in pods for vertical lift.

Two years earlier Lockheeds had tried to pre-empt the fulfilment of the requirement by offering the C-130 which was in use in the USAF; it was highly regarded but fell far short of the actual operational requirement. Harry

* I believe this figure of £7–800,000 was for a bare aircraft, the avionics being supplied by Great Britain whereas the Australians were buying a fully equipped aircraft.

Simons who was masterminding the sale and who had been successful in selling the aeroplane to many foreign countries was tireless in his efforts. I knew him well from earlier days and pulled his leg saying it was fifty to one against him selling it to the RAF. He failed at the first attempt but when the AW-681 was later cancelled the RAF bought sixty-five American Hercules C-130s instead! The lesson here was very plain; if you have a good aeroplane to sell never give it up. After all one cannot anticipate the vaguaries and tortuous planning of politicians. We tried hard to get the Tyne turbo-propellor engines installed to replace the Allison T-56 but after considerable deliberation by the Ministry they decided to stay with the T-56 engines. This I now think was the right decision as the T-56 proved to be a good reliable engine and less costly.

The TSR-2 survived a few months longer than the other two projects and then this too finally went. Much has been written about the TSR-2 and the rights and wrongs of its cancellation. My own firm conviction is that, what was more important was that it was not replaced by the Spey-engined Mirage IV but by the General Dynamics F-111 which the British Government then cancelled a year later. Politics played a decisive part in this decision; I believe it to have been ill-conceived and irreparibly damaging to our relations with the French. It had its reaction on President De Gaulle being reluctant for Britain to join the Common Market; it certainly soured relations on any further joint ventures with the French aircraft industry.

The RAF, to whom the TSR-2 had been a sacred cow for so many years, gave in with little resistance when the Government decided to cancel it: the view taken at the top was 'Who are we to argue with a political decision if we are given another aircraft which will fulfil our requirement'! The F-111 certainly should have been able to do so, had it lived up to its contract specification.

The engineers at Derby visited General Dynamics to see if the Spey engine could be fitted to the F-111 for the RAF orders; considerable engineering cost and performance studies were undertaken, a similar exercise to the one carried out for the Phantom. Meanwhile I had been lobbying to have Spey engines fitted to the Mirage IV so that it could compete and, if possible, supplant the F-111. Discussions with the operational requirements branch at the Ministry showed that the Spey Mirage would meet the requirement of the abandoned TSR-2 except for sustained supersonic speed at low altitude and for the rough field take off run. It was particularly attractive as the same reheated Spey to be used for the Phantom would go into the Mirage IV.

I invited Allan Greenwood, the Managing Director of BAC, and Jeffrey Quill (Sales Manager Military Aircraft) to a discussion in my office at Conduit Street to try and persuade them to support a proposal to share the development and production of the Mirage IV to replace the loss of the TSR-2. Allan Greenwood saw the possibilities of this and after talking it over with Sir Geoffrey Tuttle (Vice-Chairman of BAC) decided to back the scheme and approach Dassault to learn what their reaction might be.

BAC and Dassault agreed a plan and with Rolls-Royce co-operation a brochure was prepared giving performance, costs and time to production and all the other necessary information. We arranged for some RAF pilots from AEE Boscombe Down to go over to Dassault to test the aeroplane in its standard form. Their report was highly complimentary, but alas it was suppressed in the corridors of power so that the full impact was not allowed to influence the pre-judged intention to buy the American F-111.

I canvassed the support of the Opposition Minister of Aviation, Robert Carr, and other MPs of both sides of the House and also the Lords. We flew some of them to Bordeaux in a BAC aircraft to inspect the Mirage production factory. General Dynamics were invited to present the F-111 also, so that there were equal opportunities for both companies. To present the American case Roger Lewis, my friend from early days in London (when we were both trying to sell the Merlin DC4 to BOAC in 1948), came over from the States. He was now President and Chairman of General Dynamics.

From time to time the Minister of Aviation, Roy Jenkins, kept reassuring the House that the Mirage was being studied and he even postponed the decision for a while. Evidence was supplied by the BAC/Dassault/Rolls-Royce team which showed the Mirage to be able to fulfil the RAF need; it would be available earlier than the F-111 and would be much less costly. A large proportion of the cost would be in sterling rather than in dollars, it would provide jobs in Britain, it would cement Anglo/French friendship, it would find a market in other countries and so on.

I knew I was sailing rather close to the wind sometimes when I was not keeping Derby fully informed as to what I was doing, and one day I confessed to Dave Huddie, the Manging Director, that it was at times a little unorthodox. He in true Rolls-Royce spirit said, 'Do what you think is best, but if it back fires I know nothing about it.' We got a long way along the road and much closer than many people realised to getting the Mirage accepted, but the political veto was always there. The French never quite believed we could be successful and never gave their full support, although approaches were made to General de Gaulle himself.

The decision by the British Government to order the F-111 and not the Mirage IV Spey caused a cooling off in French aviation circles to working with Britain. This was serious as the next joint venture was the Anglo French Variation Geometry (AFVG) multi-role fighter, which was intended to be the replacement for many types of aircraft now in service and the intermediate ones not yet available. Dassault was already working on his own variable geometry project, so it was not long before the French opted out of the joint project and went ahead on their own. Britain then looked for other partners and so the MRCA came into being with Germany and Italy as partners.

We had questions asked in the House and through the Auspices of the Air League, on whose council I was a member, put on a presentation to the Air Committees of both Parties. This was right up their street as the purchase of the F-111 went against all their ideals. The Cabinet meeting at which a decision was to be made was set for Sunday. Having been told by Lombard, who had just returned from the States, that the F-111 development was behind schedule, that its performance was 25 per cent down and that costs were rising rapidly, I decided to make one final attempt to delay it. The problem was to persuade the Cabinet to defer its decision to order the F-111 and if possible to consider the Mirage.

On the Sunday morning I asked Air Commodore George Heycock, whom I knew from the days when he was Air Attache in Washington and later Paris, and who was now retired from the RAF and representing the French aircraft selling organisation AICMA, round to my house for a drink and to help me concoct a telegram to the Prime Minister. He naturally was keen to further the case for the Mirage and so we compiled the following message:

> 'In view of imminent decision concerning the purchase of F-111 now before Cabinet, information relative its short fall on performance of the order 25% must radically change its acceptability as a cost effective weapon. The Air League submit experienced aviation opinion would view quick decision imprudent in view of new circumstances and no action should be taken until a guarantee of performance, delivery date and cost has been re-assessed.'

Having drafted this I had to telephone round a number of other Councillors to obtain their concurrence and then persuade the Chairman, Sir Archie Hope (a battle of Britain pilot 601 Squadron AAF), to append his signature to it. He did this without demur and off it went. We heard afterwards that the telegram was delivered to the PM at a Cabinet meeting; he turned to the Minister of Defence, Mr Healey, showed him the telegram and asked if it was true; apparently he agreed so the decision was deferred until further information and guarantees were received from the States. Four days later the Government ordered the F-111. (A year or so later the deal was cancelled at a cost of many millions to the British Government.)

Having read of all these uncertainties and unpredictable selection of types of military aircraft: the reader must feel that choosing a successful project is like backing a horse in the Grand National; at times I did feel rather like a tipster. One had one's successes but there were many failures also. Some people were over confident and even arrogant about the way to go about selling engines. Such remarks as 'Rolls-Royce engines sell themselves' were never countenanced at the top level of management but it was an influence which had to be overcome at certain levels.

The Air Staff were beginning to feel that the Bristol engine company were more realistic with their future engine projects; possibly some of the Rolls-Royce ones were too far advanced in concept and not really practical, such as the engine put forward for the supersonic bomber requirement. This was a small light engine exploiting the square/cube law which promised the equivalent power of four large engines by using a larger number, perhaps sixteen, or so, at a lower installed weight. This was clearly not a practical solution and was not adopted by Avro who won the competition having decided to fit four Armstrong Siddeley A-176 engines. The Air Staff on the other hand did realise that Rolls-Royce had the ability to rectify faults very rapidly when major troubles arose. This was a valuable asset and was firmly established in their minds from the splendid performances put up during the war being a direct follow on from Lord Hives' policy of keeping the customer happy at all costs.

It was important that we should try and persuade the Air Staff of the overall superiority of Rolls-Royce, both on design ability and in achieving performance at a fixed price. In order to attract funding for new engine projects from the Ministry of Technology, they had to believe they were getting the best value for money from public funds. The best way to do this was to show the top brass what was going on at the factory in the way of advanced projects, new techniques, production facilities and last but not least expose them to the enthusiasm of the brilliant engineers, of which there were many. This exercise was done as opportunities arose; it also had the effect of letting the company executives meet the customer under friendly conditions and to appreciate their problems too.

One outstanding success came to us fortuitously about this time. This was the fitting of the Spey into the American LTV Corsair Naval strike aircraft. The aircraft had been designed to an American specification and the Pratt and Whitney TF-30 engine had been specified. My friend from earlier days Lyman Joseph was in charge of the project. He came over to England in the early stages of the design to talk to the Air Staff and to European Air Forces too in order to interest them at an early stage. I took him down to the School of Joint Air Warfare at Old Sarum to have a discussion with Air Commodore Ginger Weir and his planners who could help him and give advice from their fund of experience. It seemed rather a waste of time and effort as there was little likelihood of ever getting a Rolls-Royce engine to replace the Pratt and

The Spey Corsair

Whitney one, but he had always been kind to me in the States and so it was the least one could do in return. However, we did suggest that if ever the aeroplane was sold into Europe, perhaps the Spey could be considered as an alternative. Several years later after the Corsair had been selected by the USAF and USN it became apparent that the Pratt and Whitney engine was under-powdered and also had a dangerous surge condition which had caused some fatal accidents.

Rolls-Royce stepped smartly in and in conjunction with Allison developed a specially tailored version of the Spey to suit the Corsair and which gave considerably more thrust than the TF-30 engine, but had extremely good handling qualities also. Much credit must go to Cyril Lovesey for adapting the Spey to this installation and to Tim Kendall for looking after the commercial arrangements. It has been an outstanding success and some hundreds have been sold of which Rolls-Royce and Allison shared the reward. The success of the Spey in the Corsair greatly enhanced the reputation of Rolls in America and surely must have been an influence helping the election of the RB-211 engine at a later date by Lockheed for the Tristar airliner.

Pleasant Memories 15

Looking back on my long career with Rolls-Royce, I recall with pleasure the many projects with which I was associated, the countries I visited and the many people I met. I particularly enjoyed my visits overseas which I undertook regularly to maintain contact with our customers, particularly with the RAF, and to remind them that however far away they were, Rolls-Royce were always interested in them.

I received much generous hospitality. Most of the Commanders in Chief I had known years before when they were junior officers. In Singapore I particularly remember Hector and Jean McGregor who entertained me in Air House on my first visit to Far East Headquarters. They were followed in that same house by Sid Hughes, Peter Wykeham and finally Sir Neil and Elizabeth Wheeler who were the last people to reside there before the withdrawal from Singapore. Chris and Joan Foxley-Norris used to look after me at Cluny Road and when I stopped in Hong Kong, Dennis and Lorna Crowley-Milling used to take me to the races where they would lay on a splendid lunch in the Governor's box.

On one occasion I was invited to join the proving flight of the first VC-10 in Transport Command from Lyneham to Hong Kong by Bing Cross whom I had first known in 1933 when he was a Flying Officer in 25 Squadron at Hawkinge, flying Hawker Furies. We were to fly via Nairobi, Ghan and Singapore. We reached Nairobi safely and Doc and I spent a most enjoyable day visiting the Game Reserve. On the next leg to Ghan where we were to stop for refuelling, I facetiously asked what they would do if we had an engine failure out there in the middle of nowhere.

Bing Cross scorned the idea and told me that they had our latest Conways installed. We stopped at Ghan and after taking on fuel we prepared to take off. Three engines started but the fourth would not! It was diagnosed that the fuel control unit had mal-functioned and we would have to wait while a new one was flown out from England. I was in a hurry as I had to get to Singapore to make a connection for Australia and so the next day it was decided to fly the aircraft out with me as a crew member on three engines, leaving the passengers to stay in Ghan to await a relief Comet to pick them up. When airborne, I sent a radio message to the Rolls-Royce representative at Singapore to warn him of our problem. On arrival he met us and put the trouble right within an hour. By the time the Comet arrived the next day, the VC-10 was serviceable. It was good to know that the standard of service was being maintained abroad!

When I went to the States I often visited the British Air Attaches: Reggie Emson, Dennis Crowley-Milling, Ian Esplin and Colin Coulthard. These people, as well as being very hospitable often smoothed the way for me to obtain interviews with the senior USAF officers with whom I wished to discuss projects. I accompanied Lombard on several such occasions when he was trying to promote the multi-lift engine philosophy for V/STOL. One visit I paid to the States came about in a rather peculiar way. During the SBAC show at Farnborough, General Parker of the United States Army Air Corps called at my office in Conduit Street to ask me to tell him about V/STOL developments in Britain. I gave him my unbiased views saying that I thought for the supersonic flight I favoured the Rolls-Royce method of a mixture of separate lift engines with propulsion engines, but for his requirement in the army, the Bristol method of one engine for simplicity using vectored thrust was probably better.

A few days later I had a call from him in Saint Louis asking me to fly over to visit them at their expense to give a lecture on the subject! I got the necessary agreement to do this and flew over with Jim Heyworth, one of Lombard's assistants, to give a joint lecture. That evening, they put us on television with Mrs America. She was most attractive, but when we saw the replay afterwards, we were shocked by our appearance: we had five o'clock shadow and looked rather like a couple of burglars!

Finally, I used to visit Germany often and again I was always very well looked after and I found friends there year after year: Harry Broadhurst, Edwards-Jones, Paddy Crisham, Mike Donnett, Dickie Jones, Ronnie Lees, John Grandy and Chris Foxley-Norris. We used to go off for a day visiting the vineyards at Oppenheim for a wine tasting. Rudy Gillot had some wonderful vineyards and we got to know he and his wife very well; they gave us some wonderful wines to taste. I was usually accompanied on these visits by Athol MacIntyre, a colleague from Derby who was in the service department. We collected some superb wines and today I have just one bottle left of the Trockenbeerenauslese 1959 which can only be drunk on some special occasion. There is a tale to tell here!

When John Grandy was Commander in Chief, I gave him a bottle of this to keep. That same day at lunch, his Steward was just about to open the bottle when John noticed what it was and was able to stop him just before the bottle was opened. Later on when John left Germany to take over Bomber Command he sold his cellar to his successor Ronnie Lees. One day Ronnie was sitting in his study having some sandwiches and he asked his Steward for a bottle of wine. Ronnie was just having his first sip when the phone rang. It was John who was ringing to say that he did not include the Trockenbeerenauslese in the deal! 'Too late,' said Ronnie, 'I'm drinking it now!' He put the cork back in the bottle and bottle into the fridge where it remained until Dickie Jones and I arrived two weeks later and we all drank it together; it was like nectar. Next day we went to Oppenheim in my Continental Bentley for a wine tasting where I bought three more bottles of it.

That Christmas I arranged a dinner party at the Savoy Hotel where John Grandy was the guest. I brought one of the bottles with me and produced it at the end of the dinner. Of course, John was delighted—so he did have a taste after all!

Another time when I was in Bonn trying to persuade the Germans to accept a Rolls engine, I learned that Lord Hives was in nearby Cologne with a party of University Professors. Hives was retired and was involved with University work. I called at his hotel and was shown up to his room. He was sitting in a chair with his coat off, his feet up on another chair, with a glass of whisky in his hand, reading a book. He seemed pleased to see me and asked what I was doing over in Germany. I asked him if he would care to have dinner with me; he said he would and that he had three professors with him and did I mind if they came too? I said 'Certainly not, after all I have a rich uncle in Derby who would be delighted to pay the bill!' He laughed as he got the message—he knew that we sometimes used to call him Uncle Ernie!

These overseas visits sometimes included fishing expeditions; Adrian Lombard and I were invited to Labrador to fish the celebrated Eagle river for salmon. This was a great privilege and was due to the kindness of John Maxwell of Air Canada. It meant flying over to Montreal, then on to Goose Bay in a Vanguard. Here we met the rest of the party and we covered the rest of the journey (a distance of 130 miles) in a De Havilland Otter float plane, loaded with all our food and gear for a week. We were landed on the river where we transferred to outboard dinghies to proceed to the fishing lodge upstream; the Otter left us to return in a week's time. Lombard caught his first salmon here. At the end of the week our aircraft returned to take us back to civilisation with all our fish packed in ice. These pleasant and rare expeditions were all part of the Rolls-Royce way of life and all helped to build up a long-lasting relationship with the customer.

It was always considered to be of the utmost importance to keep in close contact with the customer, particularly when a new type of aircraft and engine was undergoing trials or being introduced into service. Prompt attention to criticisms or defects gave the operator confidence that the manufacturer had their interest at heart. The investigation of complaints promptly, even if they did appear trivial at the time carried so much more weight than if they were allowed to fester and filter through the official channels, for by then it was probably too late to prevent the product getting a bad name. Rolls-Royce had a reputation which began in the early motor car days, and which has been jealously maintained ever since for being second to none in this respect.

Once when the relative merits of Bristol and Rolls-Royce were being debated I was told by a member of the Air Council that Bristols had the more interesting engine projects but that Rolls-Royce were unique in dealing with complaints and quick to rectify faults. This policy certainly paid off as I found on one occasion in Australia. I had been suggesting to the Lockheed Electra operators that when the time came for them to replace their equipment, that in view of the structural failures they had experienced, they should consider a British replacement. Strangely I thought, they told me that as Lockheed had been so expeditious in modifying and strengthening the wings they had every confidence in that company and that if they offered a suitable replacement they would be happy to stay with them! Engineers understand the importance of this sort of service but possibly financiers may take a more shortsighted view: it can be costly at the time but pays off handsomely in the long run.

The close association and understanding between Rolls-Royce and the Royal Air Force which was built up during the reign of Hives was of great value to both. This unique understanding between manufacturer and user has been immortalised in the words beneath the Battle of Britain Window at the Company Headquarters in Derby:

> This window commemorates the Pilots of the Royal Air Force who in the Battle of Britain turned the work of our hands into the salvation of our country.

The ideals of Rolls-Royce were in essence similar to those of the RAF; namely excellence in performance, the desire for achievement, the ability to be forward looking; both had a natural flare for getting things right, not only in the technical sense, but throughout our different ways of life. The RAF still does and I felt this most strongly when I was at the last Battle of Britain Service in the Abbey. The ceremony was superbly prepared and beautifully executed. I recalled the phases that the RAF has passed through since I first knew it in 1931 when I joined the AAF: the build up and expansion to the outbreak of War, the successive changes during the war to match the prevailing strategy; the run down after victory, the change from piston engine to jets, the build up again for Korea and then the run down once more. The advent of the nuclear deterrent was the next new phase and how to achieve maximum efficiency and strike capability with drastically reduced budgets. All the changes had taken place smoothly and all the time operational efficiency and morale had been second to none. To people outside the services this side of the RAF was perhaps best exhibited by the Red Arrows and other aerobatic teams, and the winning of international bombing competitions. What a tribute the public owes to those who control the destiny of the RAF.

One is naturally very proud to have been closely associated with and to know these guardians of the Services and if one has been able to maintain a happy relationship between them, then this is a contribution which adds to the pleasure and experience of working for a company. I feel disappointed that this special relationship gradually declined under the influence of the Company's expansion into the civil market and I feel some degree of failure in my purpose of trying to maintain it.

One of the greatest pleasures of working for Rolls-Royce was meeting the many outstanding people who were associated with the aircraft and engine

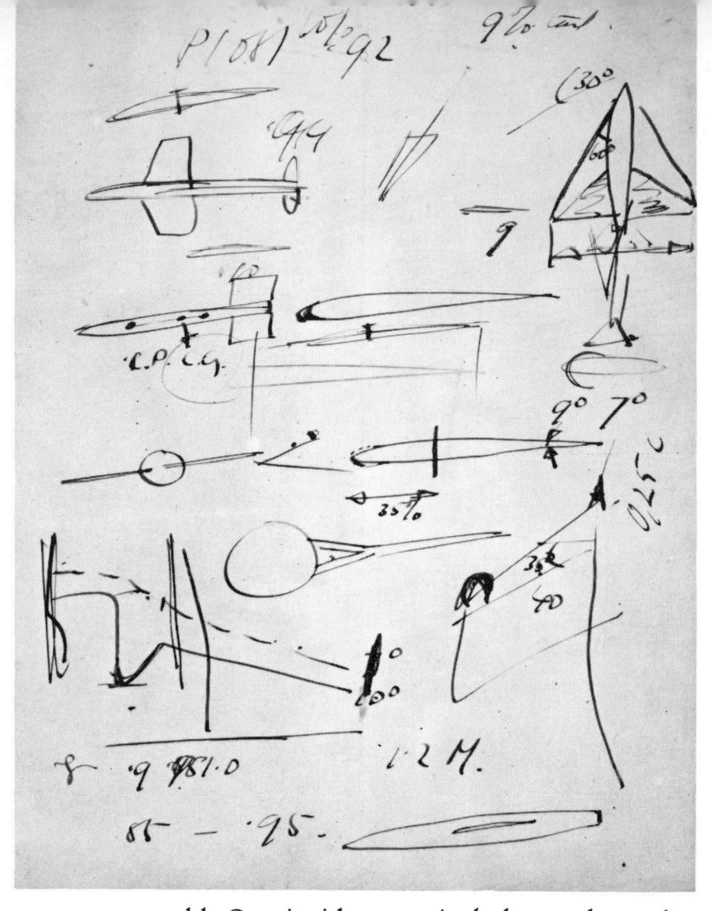

Doodles of futuristic aircraft by Kurt Tank, Sidney Camm and Joe Smith drawn on the back of a menu after a dinner organised by the author in 1951.

world. One incident particularly stands out in my mind. Shortly after the war Kurt Tank visited England to discuss post-war aircraft development. I thought it would be a marvellous opportunity to get he and some of our leading aircraft designers around a dinner table for a general discussion. I therefore invited him to have dinner at Grosvenor House and to meet Sidney Camm, who designed the Hawker fighters, Joe Smith who was in charge of design at Supermarines and so was responsible for the Spitfire development and Witold Challier from the Rolls-Royce performance department. I thought this would be highly entertaining and constructive.

The atmosphere at first was rather frigid which was not unexpected; after all they were each rather outstanding characters and could each be rather difficult at times! Tank was a Prussian, Witold was a Pole and the other two were just rather autocratic at least as far as aviation was concerned. As the dinner got under way and the ability to understand the foreign accent became easier so the atmosphere grew warmer. The Spitfire, Tempest, Mustang, FW-190 and the Me 109 battles were verbally fought and the merits of Merlin versus BMW engines were aired from several different angles. By the time coffee and liqueurs arrived designs and doodles of futuristic aircraft were being drawn on the menu. It became quite an entente cordiale and the evening ended as a memorable one for all. I can't even now get over the fact that we all met round a single table—an idea that had we thought of a few years earlier would have seemed crazy!

Multi Role Combat Aircraft 16

The types of aircraft for the re-equipment of the major Air Forces had by 1969 been decided and production was well under way. In the States it was the F-111, the Phantom and the Corsair for fighter and strike. The Phantom was finding a large export market using the American General Electric engine; the Spey engine was too expensive although it did have a better performance, but when installed in the Phantom was not as good as predicted. The F-111 had been cancelled by the RAF to obtain a suitable aeroplane to replace the cancelled TSR-2 and F-111. The requirement for these had changed somewhat since the Labour Government came to power as it had been decided to pull out of Singapore. The need for a very long ferry range had gone now that the RAF's island reinforcement policy had been abandoned; the policy was now to concentrate on supporting NATO in Europe; this required a smaller aeroplane albeit with a good low level strike capability. Negotiations had been going on with the French to produce a joint venture variable geometry strike fighter but disagreement arose between the two governments, Dassault and BAC as to who should have the design leadership; politics played an important part in the discussions and feelings ran high. The French decided to discontinue the dialogue stating that they could not afford it; it was common knowledge that they were building their own swing-wing prototype at Dassault and so the British looked upon the whole deal with suspicion.

I firmly believe that had the Spey Mirage IV gone ahead instead of the F-111 the partnership with the French would have been maintained. The Hawker-Siddeley Bucaneer which was originally built to a Royal Navy specification as a subsonic strike aircraft was now bought by the RAF and had to fill the gap until a supersonic one could be obtained. An attempt was made to form a new consortium to replace the Anglo-French one. It was hoped that Germany, Belgium and Holland and Italy would all participate and so have a common aircraft for all five countries, thus a large production would be required making it an economic proposition. The aircraft to be produced became known as the Multi Role Combat Aircraft (MRCA).

After much discussion between all the prospective members, the Belgians and the Dutch dropped out on the grounds mainly of expense, and partly because they preferred a fighter rather than an aircraft predominantly for strike. This left Germany and Italy with the UK to form a joint venture. Much discussion and controversy took place as to whether it should be single engined or a twin; this meant in turn that many engine studies should be initiated to decide whether an existing engine could be used or whether an advanced technology design which would result in a smaller and thus less costly aeroplane should be begun, although the engine development would be costly. We also had to remember that the Americans would surely try hard to produce the right engine and so keep a foothold in the European market.

In 1967 Rolls-Royce had bought up Bristol-Siddeley engines, this meant that the two companies merged and had to work together instead of against each other in the usual keen competition. There were many good reasons to justify the acquisition of the Bristol Engine Company, one of the chief ones being to form a company big enough and with sufficient experience, knowledge and talent in all fields of aero engineering to compete on the world markets with Pratt and Whitney, the highly successful American competitor. Also we had to make the best use of the rather limited financial resources of the British Government for Engine Research and Development.

It is appropriate at this point to recall my urgent report to the management expressing grave anxiety. It showed just how serious was the position on choice of engines and how keen the competition was becoming. We could easily have lost out at this point through lack of cohesion and appreciation of the competitors desire and intention to maintain the United States engine market in Europe.

Sir Denning Pearson had issued a strict instruction that the Bristol engine division and the Derby division must work closely together. Joint committees were formed to exchange ideas on costing, engineering, production, etc., to try and find the optimum joint solutions. In theory this was excellent but in practice the rivalries between teams and personalities would not die overnight and so one found situations occurring such as the marketing teams from the two divisions going separately to some of the aircraft manufacturers offering their own engine projects without agreeing a common policy from headquarters. This confusion was quickly stamped out, but it is the sort of thing that does happen when companies merge!

An advanced engine design using the three-shaft principle was put forward by Geoff Wilde from Derby and evaluated by the Bristol engine division, while the Bristol division's two-shaft engine was looked at by Derby. Eventually both divisions accepted the three-shaft engine as being the most suitable, it was then decided it should be developed as the RB-199 at Bristol. Thus the brains of the whole company were brought to bear to achieve the optimum result. After a keen and protracted battle with Pratt and Whitney and General Electric, the latter dropped out and this left us in direct competition with Pratt and Whitney again!

Internally one or two problems had arisen from the new merger, one of which nearly cost us our chance for this new contract. The problem was the internal battle for development and production time between those working on the projects for civil aviation and we who were working on military projects.

The MRCA Swing Wing aircraft

The eyes of the top management were firmly fixed on the civil market. The RB-211 engine was under development for the American wide-bodied transport market, which was being keenly contested by Lockheed and McDonnell Douglas. The engines under consideration were the General Electric CF-6 and the Rolls-Royce RB-211, both of which were large and very costly being very advanced in concept. The RB-211 was the most ambitious engine project that Rolls-Royce had ever undertaken and the most costly. Eventually the order was split. Some of the airlines ordered the McDonnell Douglas DC-10 with the General Electric engine and the others the Lockheed Tristar with the RB-211. Lockheed were keeping a tight control on the timing of the production and engine performance and were generally breathing down Derby's neck!

With no enthusiastic senior executive in charge of military projects to protect our interests, responsibility for them was then split between Bristol, who looked after the Pegasus for the Harrier, and Derby who looked after the Spey for the Phantom, Bucaneer and Nimrod. The RB-199 engine for the new MRCA project was nobody's baby yet. I was now the Military Aviation Adviser for both companies and as such I was following the progress of the MRCA very closely. Pearson asked me to co-ordinate the Company's effort to promote this engine for selection by the Anglo/German/Italian consortium in competition with the American engines. Having no executive authority it was not easy but by keeping in close touch with the Ministry of Defence and Technology, where Air Vice-Marshal Giddings and Handel Davies were so helpful, I was able to see how the chances of the RB-199 were going against the Pratt and Whitney opposition.

The Bristol engine division led by Gordon Lewis had been doing sterling work on the detailed design of the engine and the performance and cost estimation; what was lacking was the enthusiasm and drive of the Rolls-Royce Company as a whole to impress the German government of its ability to meet the performance required in the time required and for the cost that was within the budget. Visits to the US where Pratt and Whitney impressed the Germans with their resources and intention to support the programme were noted. One would also see that we were giving them the impression that we were overloaded with civil engine work and were unlikely to meet the commitment for a new and sophisticated military engine. Each country was to be awarded an equivalent amount of work to the number of aircraft ordered; the work to be split equally between the member firms in the Consortium. This meant that aircraft, engine and avionics would have to be shared also. The question of offset costs came in to the picture as well, so the Americans raised the argument of defraying the military support costs by supplying American engines. There were times when the British aircraft constructor lobbied for Britain's share to be allotted to them to enable them to compete with the engine maker. The situation was becoming serious as I realised when Sir William Cook, the British Chief Scientist who was in charge of the British negotiations, told me all that was going on. Mike Giddings reflected the RAF's fear that the choice might go to Pratt and Whitney. It was high time that Rolls-Royce should treat the matter with the urgency that it deserved and gave it the same priority as the civil RB-211. To this end I wrote a memo to Pearson, Huddie, Conway and Hinkley, concluding the following:

> 'The Germans are alone in the Steering Committee in supporting an American engine. The efforts of the Germans to bring the requirement closer to the NKF is alienating both the RAF and the Italians who are beginning to look at fallback positions in case the Consortium disintegrates. We must redouble our efforts to persuade the Germans of our superiority over the Americans on engine design, costing and ability to achieve results and to persuade the Cabinet of the necessity of supporting the British engine as one of the conditions of European collaboration. There is little time left to achieve these targets, hence the note of urgency in this memo.

Knowing what the situation was and the views of the Air Staff I arranged a dinner party at Grosvenor House for Pearson and the other senior officials of the Company to enable them to meet Sir John Grandy, the Chief of the Air Staff, Sir Peter Fletcher, the Vice-Chief, AVM Giddings, the Assistant Chief, and Air Commodore Colin Coulthard, the Director of Operational Requirements. This meeting was ostensibly for Rolls to explain why the Spey Phantom was not yet meeting its specified performance; hopefully the members of the RAF in turn would state what they felt about the chances of the RB-199 being chosen for the MRCA fighter. At the end of dinner Sir John let his hair down and told us 'The RAF would like to have a British Rolls-Royce engine in the MRCA but unless the Company showed a more aggressive spirit in promoting it the Americans would win the order.' After dinner when the guests had gone, I spoke to Pearson to see if he had got the message. He had, and right away he telephoned Hugh Conway, who was the Chief of the Bristol division, and gave him full responsibility to ensure that the Germans chose the RB-199. This did the trick and soon almost as much effort was being put into the RB-199 as on the RB-211. The Germans were invited over to see the facilities of the whole Rolls-Royce complex and the Ministry co-operated showing them the Government establishments also, thus demonstrating that Britain were in a strong position to take on the engine project in every way. Shortly after this agreement was reached, and the RB-199 was chosen!

It is amusing to recall the lighter side of this rather important dinner. I had gone to much trouble to arrange the menu with Bob Sawyer the Managing Director of Grosvenor House; we had chosen 'Saumon Kubliak' with 'Le Montrachet' 1964 to go with it; they only had a dozen bottles left. The dinner was superb, but the atmosphere had been rather intense and this splendid fare had gone rather unnoticed to my disappointment until afterwards when I accompanied Sir John to his car, he remarked how good it had been and the wine so very special. It was only two years before when he was Commander in Chief in Germany and he had nearly missed tasting the Trockenbeerenauslese of 1959.

The internal problems, however, had not been so easily resolved and during 1960 it became increasingly clear that the RB-211 engine was again absorbing most of the attention of the top management, and that other projects were suffering as a result. I was particularly worried about my military customers as they were telling me in no uncertain terms what they thought about the poor service and attention they were getting. There was a shortage of engine spares on all engines, the overhaul schedules were not being met and the aircraft were being grounded as a result.

In my capacity as Military Aviation Adviser I felt that the customer had a very genuine grouse; it was very frustrating for me to be unable to get the message across or to get anything concrete done about it. To help the situation, I instituted a committee, agreed by the Chief Executive, called the Military Aviation Year Committee. The object of this was to highlight the importance of the military customer, to bring their complaints to the notice of all divisions of the company, and to try and get something done about them. The most important customer in this context was, of course, the RAF, who in overhauls and spares alone paid the firm over £40 million per year. I also formed a Future Fighter Committee, the object being to review world-wide fighter requirements and advise the engineering departments on what kind of engines would be needed to suit the various military customers. I hoped that the recommendations of these two committees, which consisted of Sammy Wroath, Frank Brittain, Witold Challier and other members from all three engine divisions would bring to the notice of the Company Administration the importance of treating military aviation projects as seriously as the civil. I also hoped to be able to co-ordinate the views of each division so that there would not be duplication.

A series of visits was arranged to each Operational Command of the RAF at which representatives of each division (engineering, service and test pilots) were lead by the Managing Directors of each of the divisions concerned so that the Commanders in Chief of the Commands would be able to air their complaints directly and at top level. These visits were of value and there was some very plain speaking by the Command Staffs, notably Air Vice-Marshal Mick Pringle, Senior Officer at Bomber Command. The results of these meetings were however less impressive; promises were made to improve the supply of spares and attention to their complaints were put on higher priority, but the system was too much in arrears for any quick results to be obtained. The effort required in trying to meet Lockheed's target with the RB-211 was occupying too much of everybody's time.

As a final gesture I received agreement from Pearson to hold a dinner party for the Air Board so that there could be a frank exchange of views between them and the executive Board Members of the three Rolls-Royce engine divisions. I thought that this would really bring the message home (as it had at the previous dinner at which the Chief of Air Staff had read the riot act on the RB-199 for the MRCA), and that there would be an improvement. We spent a very pleasant evening at the Savoy; we had all been to Farnborough Air Show and were therefore rather tired and so it stayed on a pleasant social level. Chief of Air Staff Sir John Grandy told some amusing stories and that was as far as we got! Perhaps it was felt by our guests inappropriate to talk shop on this occasion and little reference was made to their complaints. However, after the guests had left, Sir Denning spoke sternly to the Executives present and said there must be an improvement in customer relations and better service given with higher priority in dealing with the complaints of the RAF. Thus the message had been received; sadly it proved to be too late.

So Rolls-Royce after a hard struggle won these two vital contracts. The RB-211 ensured the future in the civil market; the RB-199 which will be the advanced European military engine of the next decades, was won for the military market.

Confusion 17

The other major problem concerning the RB-211, apart from the amount of development time it was absorbing, was the finance; it was the most costly project that Rolls-Royce had ever undertaken and of course it was draining finance from other projects. These internal problems were becoming acute, but of course one had to keep working and attending to the immediate needs of one's customers. I now decided to look at the people who had opted out of the MRCA competition—the Dutch and Belgians as they had yet to select an aeroplane. I also concerned myself with Australia and America.

The Royal Australian Air Force had just issued a requirement for an advanced supersonic fighter to replace their Mirages. In their case and in the case of the Belgians and Dutch, I felt that the most important thing was to keep the cost down and that it would therefore be too expensive to design and build a completely new aircraft for them. I felt that the considerable increase in performance which they required could be achieved by adding engine thrust to their existing airframes, which would have to be suitably modified to accept it.

There was a development of the Bristol Viper with reheat likely to be available which if added to either the Mirage and Lockheed 104 would give 30 per cent increase in performance and at the same time add a safety factor; if the main engine failed the Viper would 'get you home'. The engine was to be installed on top of the fuselage at the rear just like the Tristar or the Trident. The Bristol Division did some studies on these lines and produced a brochure for me to take to the Continent and to Australia. This I did, but with no success as they feared that the behaviour in flight might be impaired. I returned in November 1970 to make a final bid to get a Rolls-Royce engine accepted but again to no avail. It is interesting to know that six years later as I write this book there has still been no decision in Australia on which aircraft to purchase and the Dutch and Belgians have only recently ordered the General Dynamics F-16!

I next visited the Pentagon and saw General Glasser in charge of requirements for the USAF and asked him if he would consider a Spey development of the TF-41 Anglo-US variety reheated giving 26,000 lb thrust which would be competitive with the new Pratt and Whitney F-100 engine. He was quite willing to assess any proposal we might like to put forward particularly if it was likely to be less costly. I was sure it would be as it was to be a development of an existing engine whereas the Pratt and Whitney one was completely new.

When I returned home I found the firm in a state of impending crisis and I was told that there would be no chance of raising money for its development. This was a pity, as now six years later, the world market for the engine is huge, much bigger than the RB-199. I also found that rumours were rife that finance had dried up, that the RB-211 was behind schedule and that it was over the cost-estimate down on performance and overweight. In spite of this the company was trying to raise more funds to design and develop a larger and improved version of the engine, the RB-211/56, for the BAC-311 and a long range version of the Tristar. This seemed illogical to some of us and we felt that this engine would break the company (never for one minute thinking that the present versions could!).

It was at this point that Sir Denning Pearson stepped down from the Chair but with promise from the Government of an injection of £47 million to keep the project alive.

After Sir Denning's departure from command, the morale of all levels of staff deteriorated rapidly; the two main divisions which had been coming together gradually after the take over sprang apart and the old jealousies reappeared. Derby was the target to be humiliated by the new management. This, in spite of their past tremendous record. There were elements, who at this late stage wished to belittle the RB-211, just when it needed full support and collaboration from the whole company. However from the technical standpoint a prudent move had been taken, which was to strengthen the team by bringing in some of the old experienced engineers who had developed the world famous Merlin from the other divisions to which they had been dispersed, to expedite the development of the RB-211; so an all out effort was being made to get the engine right as quickly as possible. In actual fact, the engine did a successful run on the test bed, meeting its first objective on the day the Company was declared bankrupt. Too late to alter the decision, it was rather reminiscent of the Skybolt missile having a satisfactory firing when President Kennedy scrapped it at the Nassau meeting. If only it had been possible to achieve the required performance a little earlier or if Adrian Lombard had not died prematurely, who can tell what the outcome might have been!

What a tragedy it was that Pearson's leadership was terminated at this critical juncture; it broke the link with the traditions of the past and left the field open to the new management to manipulate the control at the headquarters and to exploit the difference between the three aero divisions as they wished, without the experience which was so necessary in dealing with such complex and sensitive an organisation. Confidence in the management and its aims dwindled, and esprit de corps broke down. Juniors were asked what they thought of their bosses and seniors were asked their opinions of their juniors! Thus people became apprehensive about their futures at a time when it was inevitable that there were going to be redundancies. Senior and valuable executives started looking elsewhere for other jobs and there was a danger of a massive brain drain.

Many of these people who were well known by the various customers throughout the world and who had known the customers, their needs and requirements, left the company; thus a link forged over the years at great cost of irreplaceable experience was to be lost to the new nationalised company. One frequently heard our customers, both military and civil saying that they didn't know who the new executives were and that they were damn sure they didn't know who they were!

Meanwhile the new company had instituted an economy drive: restrictions were placed on travel, newspapers were cancelled, tea was given in paper cups! and the Directors' Bentleys were sold and they were given Rovers and Triumphs. I had recently had my Continental Bentley repainted having reached agreement with the administration that I could purchase it on my retirement. This was now taken over by Hugh Conway, the new Senior Executive, but he was too long in the leg for it to be comfortable and so it was sold! I complained but to no avail; I was offered a Rover instead but this never materialised. Things were getting bad!

I held what was to be the last of my 'Future Fighter' meetings which was attended by the new Chief Executive (engineering). We discussed the MRCA project which was now well underway in the 'project definition' stage incorporating the RB-199 engine. It was decided, quite rightly that this should be number one priority and that no support should be given to another project which had been proposed; a single RB-199 engined single-seat air superiority fighter to be complimentary to MRCA.

There was another project however which I strongly advocated should be a joint development between Allison and Rolls-Royce. This was the TF-41 engine which was so successful in the American LTV Corsair. If it could be fitted with an afterburner it would give 26,000 lb thrust. This would then be an alternative to the Pratt and Whitney F-100 which was being designed for the New American fighters; the Grumman F-14 and the McDonnell F-15. The Rolls-Royce engine would be cheaper, available earlier but perhaps a little heavier, but would be a good alternative and insurance policy if the Pratt and Whitney engine was late or had serious troubles to overcome. I had been told by General Glasser at the Pentagon during my visit to Washington that he would give it consideration, if he was supplied with a brochure and all relevant data.

Alas my proposal was overruled by the new Chief Executive, on the grounds that all effort should be devoted to the RB-199 engine and that the TF-41 might interfere with its progress. It was hard to see why it should as the RB-199 was being developed at Bristol and the TF-41 would be developed jointly by Derby and Allison in USA. No effort was therefore made to obtain funding either from the US or British governments and so the project did not proceed. The prize would have been great and might well have been attained. The Pratt and Whitney engine was late, it did have troubles, its cost did escalate greatly; so had the TF-41 been available it would have stood a very good chance of at least sharing the market. The numbers ordered to meet the

US aircraft programme have greatly exceeded the numbers of RB-199s ordered for the MRCA. After the meeting I wrote the minutes inferring that further consideration would be given to the TF-41 project but was instructed to delete this and say work should not proceed; and so a great opportunity was lost.

A few days later I was summoned by the new Chairman, Lord Coles and asked what my position was in the company. I duly explained that I had been with the firm for over forty years etc., and that my main interest was to look after the requirements of the Royal Air Force. He asked me if they were happy! So I told him, firmly and clearly that they were not. He asked me what I was going to do about it saying that where he had just come from (Unilever) the customer was always right. I remarked that it used to be so too at Rolls-Royce. Answering his question, I suggested that he should make it high priority in the Company's policy to concentrate on the RAF in an equal manner to that shown to Lockheeds with the Tristar. He then asked me 'What would the RAF say when they heard that the larger version of the RB-211 was not going ahead' (he had just heard that morning that it was not going to be supported). I said 'They would be delighted'; this again surprised him, as he said he had been told by his staff that they wanted the engine to be installed in the Lockheed C-5A which they contemplated ordering for Air Support Command! This was so far from reality it shows what wishful thinking can do. He then said 'Thank you, Mr Harker, I am interested in what you have told me, I am going to a board meeting this afternoon.'

Shortly after this, the Director of Personnel asked me if I would like to retire early as the Company was in a poor way financially and was cutting back drastically. I replied that I would certainly not. About a week later (15 January) he told me I would have to go on 1 March and offered me a lump sum equivalent to my remaining contract. A few days later after consulting my lawyer and accountant, I was advised to accept. Things were really moving fast by now. On the 27th, I was promised a cheque in lieu of the broken contract, but it was never signed and on 1 February all payments were suspended, so I never collected. Even the pension fund cheque was returned but later honoured. The Receiver some weeks later to a certain extent compensated by arranging an increase in pension but the cash in lieu of the unexpired contract never materialised. It was all rather traumatic and unbelievable; but when a company is going on the rocks, things do happen very quickly and panic sets in. The whole aviation world, indeed the whole international business world was shocked that this could happen to Rolls-Royce.

On 4 February 1971 the Receiver was called in as the Company was bankrupt and so ended Rolls-Royce Ltd after sixty-five years of unparalleled fame and sound trading.

When the caretaker management took over after Sir Denning Pearson stood down things changed; close personal relationships vanished overnight or at least went into hibernation perhaps to be reawakened under new control which will be exercised by the Government and a new board. Leadership was needed to revive the old spirit; the talent remained dormant and only needed stimulation. It was analogous to a squadron in war suffering heavy casualties; the squadron must withdraw for a rest and to reform with a new and acceptable leader, train up and go back into the line. New leaders are hard to find; surely Rolls-Royce deserves the best and surely one will emerge of sufficient stature, charm and personality, coupled with a depth of knowledge of the aircraft industry to take the helm and lead the company back to its rightful place as world leader in engineering. There is nothing wrong with the middle management which produces the work and which has the traditions deeply embedded. It only has to be stimulated and there will be nothing to prevent a return to the former position of high regard and happy relationship with the customers and operators. A way of life can change under strong pressures, but once a code of behaviour has been established it is not likely to alter in a lifetime. The Rolls-Royce tradition will remain and will leave its mark for the future.

The Government came to the rescue and nationalised the divisions of the company which produced defence equipment in order to maintain the flow of engines for the services. The Receiver sold the Motor Car Division back to the public who wished to buy shares. It now flourishes as Rolls-Royce Motors Ltd, a public company worthily upholding the 'Magic of a Name'.

After nationalisation, the Aero and Marine divisions, now known as Rolls-Royce (1971) Ltd, moved its headquarters to London. This was exactly opposite to the action Hives had taken when he became General Manager back in 1936. He had then moved the authority from the London Office back to Derby so that major policy and engineering decisions could be taken at the factory where the engineers contracts people and accountants worked and where he presided with his co-directors of engineering, finance and marketing, etc. An adequate liaison was maintained in London to keep in close touch with the various ministries to discuss operational requirements and other matters of importance. But the dynamo was the factory from where the energy flowed. Only time will tell how this reversal of administration will work out and things may be different now when finance seems to be the most important consideration, rather than concentrating on the pure needs of the customer. Several years have now gone by and it is gratifying to be able to say that the present leader is upholding the traditions and dignity of the old company. I was recently invited to revisit the Derby works and was pleased to learn from old colleagues who were still there that they were in good heart.

Epilogue

The Board of Trade has published its report and formed its own conclusions as to what went wrong at Rolls-Royce. The report was unstinting in its criticism of the Chairman and the Managing Director of the Derby Aero Division, but it quite clearly missed aportioning any blame to the other directors, a number of whom to my knowledge, knew an awful lot of what was going on with the RB-211.

It is not possible in a firm like Rolls, where frank and constructive discussion takes place between senior executives, to keep opinions secret. It had become common knowledge that something was dreadfully wrong; it had been guessed in the Industry generally and British and Foreign aircraft firms had not been slow to hint at their misgivings. The Bristol division had been highly critical of the progress of the RB-211 although they had troubles of their own with the problems on the Olympus for the Concorde escalating in a similar fashion to the RB-211. There was a subtle difference in the two cases however as the Concorde was Anglo-French and so was inviolate! The Pegasus also was causing some headaches, so both divisions had had their work cut out to satisfy the customer. It is hard to imagine that discussion at Board Level did not throw up the seriousness of the general position, particularly as both divisions were represented on the main Board. Let me quote *Air Pictorial* in an article discussing the Board of Trade report on the bankruptcy of the Company.

> 'In the report there is apparently severe criticism of two men who have rendered great service to British Aviation and to world communication —Sir Denning Pearson and Sir David Huddie. The Department of Trade and Industry did not have the courtesy or even common good manners to send them copies of the report—they must pay £5 a head to read criticism of themselves—and they are reported as saying that they were not told in advance that they were being singled out for special criticism. This is typical Government action and in Britain all governments have an equally bad record. It is not really too exaggerated to say that if a man in Britain gives a lifetime's service to aviation he will sooner or later be discarded in the dirtiest way possible and replaced by someone who boasts of his lack of knowledge

of the business—just look at our present set up.

'The main criticism of the top Rolls-Royce men appears to be that they were engineers and not accountants. Thank God they were engineers. It has been the engineers who have saved this Country while the accountants have done their best to wreck it.

'Let us look briefly at the achievement of Rolls-Royce. In World War 1 the Eagle engine made its contribution and it was this engine that made possible the first non-stop aeroplane crossing of the North Atlantic and the first flight to Australia. Eagles powered many of the first transport aircraft—not only British ones—and Eagles, Falcons and Condors played a major role in the early years of the Royal Air Force. The Rolls-Royce 'R' engine won the Schneider Trophy, an achievement which the accountants did their best to prevent and the Kestrel paved the way for the Merlin which saved the country in 1940. I think I am correct in saying that every single engined fighter used by the RAF in the Battle of Britain had a Merlin engine.

'It was Rolls-Royce that gave us the Dart propeller-Turbine. Surely this was a gamble which no commercial company should have been involved in because no airline or air force was using propeller-turbines. The accountants should have stopped that. But the Dart made possible the Viscount which achieved more sales than any other British aeroplane of comparable size and it powered the Fokker F-27 Friendship which is the world's best selling aeroplane in its category—well over a thousand Viscounts and F-27s which represents about 3,000 engines without allowing for spares and replacements.

'The Avon was an outstanding success in the civil and military fields and the Spey hasn't done too badly either. The Conway was the first of the bye-pass engines to enter service and we must not forget the achievements of Rolls-Royce in developing direct life engines. The RB-211 was a gamble certainly, but the men at Rolls-Royce knew that the new technology engine was the power plant of the next decades. The Boeing 747, the DC-10, Tristar and the A300 Airbus were all powered by the new generation of big fan engines. Apart from their performance these engines are the first to bring relief to those on the ground, these engines are quiet and the RB-211 is the quietest of them all.

'Noisy engines will no longer be tolerated and every new aeroplane will have to be capable of meeting ever more restrictive noise rules. Therefore, it is obvious that if Rolls-Royce had not built the RB-211 or its equivalent, Britain would have ceased to be a supplier of transport aircraft engines at a critical period in the development of air transport. We do not know how far the government was pushing Rolls-Royce to get the Tristar engine order.

'The leaders of Rolls-Royce are accused of making wrong decisions leading to the bankruptcy of their company. This is an interesting finding in a government report when the same government is obviously determined to go ahead with Maplin against massive advice that the project is wrong—a government that is equally determined to construct the channel tunnel; two projects which are likely in combination to bankrupt the country and certain to deprive it of urgently needed development in other regions; only time will tell.'

(The subsequent socialist government cancelled both these projects.) It is idle to speculate now what might have happened had the government taken a different course and been willing to do an Upper Clyde, British Leyland or Chrysler operation, but nevertheless it is interesting to do so! Lockheeds could not have existed without Rolls-Royce, so perhaps the penalty clause would not have been invoked; it was limited to £20 million anyhow.

To put it in fashionable terms, Rolls-Royce had suffered a cash-flow problem. It had come first to Rolls-Royce because they were always funding long term, ambitious projects in the forefront of technology: a laudable policy but one which in these last years finally beat them because they had not taken into account the escalating world inflation, due to Arab oil monopoly, rising prices, high interest rates and a general trade recession.

Since then numerous other aero-space companies and even giant motor corporations, who spend little on research and development, have been in grave trouble requiring massive government financial assistance. In retrospect it seems unfair that Sir Denning was deposed when his successors and many other companies find themselves in the same position. Perhaps though this was the price of being a pathfinder. Many observers believe that had the chosen spokesmen who reported to the Prime Minister just prior to the fateful decision given a more accurate instead of an over-pessimistic assessment of the situation; or had Lord Beeching's advice been heeded, when he said there was a fair prospect of the company being saved by a further injection of capital, believing that the RB-211 would become profitable, the company could still have been in the hands of the share holders today. Since then more than 500 engines have been sold, it is the quietest of the three large competitive civil engines, as well as being the most economical and reliable, and many more are on order. A handsome profit of some £50 million has been made against a forecast loss of £45 million!

I cannot speak from personal experience what has happened after 4 February 1971, which is the date I left the Company. One can only hope that this great company of wonderful people will rise to its pre-eminence and former glory, but it can only do this by leadership based on the precepts of the great men of understanding in the past who have guided its destiny through two world wars, and made the name a household word synonymous with excellency and good fellowship. As this book goes to press there are encouraging signs that Rolls-Royce (1971) Ltd is making good progress under its new Chairman to whom one wishes good fortune.

On 4 February 1971 the time came to say 'Goodbye' and to clear out my accumulation of papers which went back many years. My secretary, Kathy Harris, decided to leave too, although she had been offered a job with one of the new directors. I received kind letters from my many friends in the RAF and the industry regretting the break in close relations with the firm. My wife and I were entertained by several of the RAF and industry chiefs, notably Air Chief Marshals Sir Neil Wheeler and Sir Dennis Smallwood; this kindness did much to soften what after all was a traumatic and bitter experience.

My 'signing on' at the Labour Exchange and collection of the 'dole' was unpleasant and had it been necessary for a long time would have been depressing. It reminded me of the first time I had been laid off forty-one years before—but of course this time there was no Hives to call me back to work on the Schneider Trophy!

I was soon engaged on consultancy work and writing this book. Working for one's self in spite of the socialist governments penalties for doing so has its compensations; to be master of one's own time does have its attractions. Working for a number of companies also broadens one's outlook, making one become acquainted with new activities, personalities and situations and forcing one to be self reliant.

In retrospect, the memories of Rolls-Royce are rich and unforgettable. It was a privilege to be associated with such a fine company and so many of its interesting personalities from whom I have learned much of value in life. It also brought me into close contact with so many splendid people in the Services; for this I shall be eternally grateful.

I can only hope that the spirit of the company imbued by its Founders, pioneers and all workers will continue and that the company under the leadership it deserves will in due time regain its former glory and affection of its customers and associates.

Rolls-Royce Engine Installations 1925–71

Engine	Installations
Eagle VIII	Phoenix Cork
	Felixtowe F.3, F.5
	De Havilland 4, 9A, 9B, 10B, 10C
	Fairey IIIC, IIID
	Handley Page 0/400, V1500
	Short Shirl
	Vickers Vimy, Vernon
Hawk	Sage Trainer Type III
	B.E.2E
	Avro 504F
Falcon	Bristol Fighter F.2B
	Martinsyde F.1, F.3, F.4, R.E.6, 7
	Armstrong Whitworth F.K.12
	Avro 523C, 529 Pike
	Blackburn S.P., G.P., Kangaroo, Spratt
	D.H.4
	Fairey F.2
	Sopwith tractor triplane
	Parnell Perch
	Short 184
	Vickers Vendace
Kestrel	Avro Antelope
	Blackburn Nautilus, Sydney
	Fairey Fox I, II, III, IV, IIIF
	Firefly II, III, Fleetwing, Hendon II, S.9/30
	Gloster C.16/28, Gnatsnapper III
	Handley Page Heyford
	Hawker Hart, Demon, Fury, Osprey, Nimrod, Audax, High Speed Fury, Hardy, Hartbeeste, Hind
	Miles Master
	Parnall Pipit
	Saunders-Roe A.10
	Short Singapore II, III, Gurnard
	Supermarine Southampton
	Vickers 141, B.19/27, 163
	Westland Wizard
Buzzard	Blackburn Iris V, VI, M.1/30, M.1/30A
	Handley Page H.P.46
	Hawker Horsley
	Short Singapore 1, K.F.1 Sarafand
	Vickers M.1/30
Goshawk	Blackburn F.7/30
	Bristol F.7/30
	Gloster S.15/33
	Hawker High Speed Fury II, P.V.3
	Short R.24/31
	Supermarine and Westland F.7/30
	Westland Pterodactyl V
Merlin II, III	Supermarine Spitfire I
	Boulton Paul Defiant I
	Hawker Hurricane I, Sea Hurricane I
	Fairey Battle I
IV	Armstrong Whitworth Whitley IV
VIII	Fairey Fulmar I
X	Handley Page Halifax I
	Vickers Wellington II
	Armstrong Whitworth Whitley V, VII
XII	Supermarine Spitfire II
XX	Bristol Beaufighter II
	Boulton Paul Defiant II
	Handley Page Halifax II, V
	Hawker Hurricane II, IV
	Avro Lancaster I, III
21	De Havilland Mosquito I, II, III, IV, VI
22	Avro Lancaster I, III, York I
23	De Havilland Mosquito I, II, IV, VI, XII, XIII
24	Avro Lancaster I, III, York I
25	De Havilland Mosquito VI, XIX
28	Avro Lancaster I, III, Kitty Hawk II
29	Hawker Canadian-built Hurricanes
	Curtis Kittyhawk II
30	Fairey Barracuda I, Fulmar II
31	De Havilland Mosquito XX
	Australian Mosquito 40
	Curtis Kittyhawk II
32	Fairey Barracuda II
	Supermarine Seafire II

33	De Havilland Mosquito XX, 40, 38
35	Boulton Paul T2
45	Supermarine Spitfire V, P.R.IV, VII
	Seafire II

45M	Spitfire L.F.V
46	Spitfire V, P.R.IV, VII, Seafire I
47	Spitfire VI
50	Spitfire V
50M	Spitfire L.F.V
55	Spitfire V, Seafire III
55M	Spitfire L.F.V, Seafire L.F.III
60	Vickers Wellington VI
61	Supermarine Spitfire VII, VIII, IX, P.R.XI
62	Vickers Wellington VI
63	Supermarine Spitfire VII, VIII, IX, P.R.XI
64	Spitfire VII
66	Spitfire L.F.VIII
67	North American Mustang III
69	Mustang III, IV
70	Supermarine Spitfire H.F. VIII, IX, P.R.XI
71	Spitfire H.F.VII
72	De Havilland Mosquito XVI
	Westland Welkin I
85	Avro Lancaster VI, Lincoln I
224	Lancaster I, III
225	De Havilland Mosquito 25, 26
266	Supermarine Spitfire L.F.XVI
113, 114	De Havilland Mosquitos 34, 35, 36
130, 131	De Havilland Hornet
500/20	Fiat 659

Peregrine	Westland Whirlwind
	Gloster F.9/37
Griffon	
II, III, IV	Fairey Firefly I, II
	Supermarine Spitfire XII
VI	Seafire XV, XVII
61	Spitfire 21
65	Spitfire XIV
66	Spitfire XIX
64	Spitfire 21, Seafire 46
67	Spitfire XIV
69	Vickers-Supermarine Spiteful
72, 74	Faireys Prototype Firefly IV, etc
85	Supermarine Spitfire XIV, 21
	Seafire 45
85, 87, 88	Spitfire XIV, 21, Seafire 45, 47
37	Fairey Barracuda V
67	Avro Shackleton MR2
57A	Shackleton MR3
59	Fairey Firefly A57, A58
Exe	Fairey Battle Test Bed
Vulture	Avro Manchester
	Hawker Tornado
	Vickers-Armstrong Warwick
Eagle The Second	
22	Westland Wyvern T.F.1
Welland	Gloster F.9/40, Meteor
Derwent	
I, II, III,	Gloster Meteor
IV, 5, 8,	Meteor
8	Meteor, T7, F8, FR10
	Fokker S.14
9	Armstrong Whitworth Meteor
	Short S.B.5
Nene	
3	Supermarine Attacker F.1
Mk 10	Canadair Silver Star Mk 101
Mk 101	Hawker Seahawk
Mk 102	Supermarine Attacker F.B.2
	Gloster E.1/44
	Fokker S-14
	Boulton Paul PIII A
	Dassault Ouragan
103	Armstrong Whitworth Sea Hawk 100
RN 2-21	Fiat G.82

Tay	
J48	Lockheed F-94C
	Grumman F9F-5 Panther, F9F-6 Cougar
Verdon	Dassault Mystere II, III, IV, IVA
Trent	Gloster Meteor Test Bed
Clyde	Westland Wyvern
Avon	
AJ.65	Avro Lancaster Test Bed
	Gloster Meteor
R.A.3	English Electric Canberra B2, PR3
	B5, T4, B6, PR7
R.A.7	De Havilland 110
	Dassault Mirage 111 O, Entendard
	Commonwealth CA-27 Sabre
	Short S.A.4
R.A.7R	Dassault Super Mystere B1
	Supermarine Swift F4, F.R.5
	Saab 32 Lansen, Draken
R.A.14	Vickers Armstrong Valient B1
R.A.21	Hawker Hunter F1, F4, T7
101	English Electric Canberra T11
109	B.1.8
115	Hawker Hunter T7
117/8	De Havilland Comet 2C
200 series	F6
	Fairey Delta 2
	BAC Lightning F1, T4
	Hawker Siddeley Sea Vixen
202	Vickers Supermarine Scimitar
204	Valiant
207	Hawker Hunter G.A.9, F.R.10
206	English Electric Canberra PR9
210	Lightning F.2
26	Commonwealth CA-32 Sabre

R.A.29	De Havilland Comet 4
	Blackburn Bucaneer I
Dart	
514	Boulton Paul Balliol
511	Fokker F27
101	Armstrong Whitworth Argosy
	Breguet Alize
	Andover
	Grumman T6 4C

Tyne	Breguet Atlantic
	Canadair Yukon
	Short Belfast
	Transall C 160
	Fairey Rotodyne
RB 108	Balzac
	Short SC-1
RB 145	Dornier VJ-101
RB 162	Mirage 3V
	Dornier DO-31
RB 193	Vak 191
Conway	Handley Page Victor B2
	Vickers VC 10
Spey	Hawker Siddeley Nimrod
	Bucaneer 5.2
	McDonnell Douglas Phantom II
	Chance Vought A.7

Adour	SEPECAT Jaguar
	Hawker Siddeley Hawk
	Japan T.2

The 19th hole at Wentworth. From the left are: Sir George Edwards, Author, Sir Hugh Kilner, W. Gill and Chapman.

The author, Roger Lewis and Huddie at the Paris Air Show

Lombard and Harry Pearson at the SBAC Show, Farnborough

Sam Patch, Fred Rosier, Tommy Thompson and Bill Horrocks